S0-BPZ-020

Automation in Animal Development

A New Theory Derived from the Concept of Cell Sociology

Monographs in Developmental Biology

Vol. 16

S. Karger · Basel · München · Paris · London · New York · Tokyo · Sydney

Automation in Animal Development

A New Theory Derived from the Concept of Cell Sociology

Rosine Chandebois

Laboratory of Animal Morphogenetics, University of Provence, Marseille, France

In collaboration with

Jacob Faber

Hubrecht Laboratory, International Embryological Institute, Utrecht, The Netherlands

33 figures, 1983

S. Karger · Basel · München · Paris · London · New York · Tokyo · Sydney

Monographs in Developmental Biology

National Library of Medicine, Cataloging in Publication
Chandebois, Rosine
Automation in animal development: a new theory derived from the concept of cell sociology
Rosine Chandebois, in collaboration with Jacob Faber. – Basel, New York, Karger, 1983.
(Monographs in developmental biology; v. 16)
1. Cell Differentiation 2. Gene Expression Regulation 3. Embryology 4. Morphogenesis
I. Faber, Jacob II. Title III. Series
W1 MO567L v. 16 [QH 491 C454a]
ISBN 3–8055–3666–6

Drug Dosage

The authors and the publisher have exerted every effort to ensure that drug selection and dosage set forth in this text are in accord with current recommendations and practice at the time of publication. However, in view of ongoing research, changes in government regulations, and the constant flow of information relating to drug therapy and drug reactions, the reader is urged to check the package insert for each drug for any change in indications and dosage and for added warnings and precautions. This is particularly important when the recommended agent is a new and/or infrequently employed drug.

Printed in Switzerland by Thür AG Offsetdruck, Pratteln
ISBN 3–8055–3666–6

Contents

Preface

Over the last 30 years there have been enormous strides made in the study of developmental biology, due in part to the parallel growth of the science of molecular biology following the discovery of the rôle of the DNA molecule in the replication and functioning of the genetic material. Such was the impetus of this new science that many came to believe that it subsumed the science of developmental biology as part of itself: DNA provided in encoded form instructions for the development of the organism, to be implemented through the control of protein synthesis. To account for the emergence of shape and form in the embryos, and especially their mysterious capacity for regulation, some biologists, and a welcome influx of mathematicians and computer scientists, turned to the study of pattern formation. The most popular models have invoked gradients of supposed morphogens across embryonic fields: gradients, plus sequential gene derepression, seemed to many to provide explanation enough.

Prof. *Chandebois* believes that such a simple scheme is inadequate to account for all the properties of the developing embryo and, in a series of important papers over recent years, she has developed a model which seems truer to the facts of development. In this book she presents her model, with the distinguished assistance of Dr. *Faber*, in the context of a wide-ranging review of the literature, extending from the classical studies of experimental embryology to the most recent discoveries of molecular biology. The model, which with the aid of excellent illustrations is described with admirable clarity, suggests an explanatory scheme that covers all levels of embryonic organisation, from molecular to morphogenetic, in a way which does justice to all of them. Much emphasis is thrown on the cells, and especially on the elementary patterns of social behaviour which characterise them and which are a vital feature of development. Gradients arise as secondary features, arising from interactions between cells rather than as supracellular phenomena which determine the differences between cells. The whole cell, not simply its chromosomal material, is regarded as the repository of developmental information; indeed, the DNA is regarded in this model as analogous to the arithmetical circuits of a computer, processing information supplied to it from the cytoplasm. Development is the fulfilment of a programme contained in the egg, but not exclusively in its DNA.

The appearance of this book is timely, for there is a growing awareness that current explanations of development require re-examination, a growing attentiveness to the part played by the egg cytoplasm in controlling development, and an enhanced vision – largely the gift of the scanning electron microscope technique – of the relation of cells to each other and to their extracellular environment within the embryo. It will be read with great interest by all who are concerned with the most basic problems of developmental biology, and I believe that its influence will be considerable.

D.A. Ede, Glasgow

Foreword

The direct observation of a natural phenomenon or its analysis, even with the most sophisticated techniques, never leads to absolute certainty. It only provides indications, which would remain without significance, could they not be brought into connection with other indications assembled and interpreted earlier by others, sometimes indeed long ago. Science ought to proceed in the manner a detective conducts an investigation: each time a new indication does not tally with what he had deduced from those already available, he goes back and reconsiders all the indications assembled so far. For the biologist the indications are unfortunately scattered in thousands of documents. Moreover, the beginnings of the investigation are to be found in his cultural heritage. Just as a legend in the course of time tends to merge with the event that gave rise to it, a theory that is transmitted by cultural inheritance tends to be transformed into firm evidence. Thus, new indications no longer call the theory into question. On the contrary, the theory is often used as an argument to cast doubt on certain techniques or to eliminate an 'embarrassing' result, much as natural selection eliminates badly adapted forms. In a limited field of enquiry a theory that was wrong from the start may still seem to be proved beyond doubt. However, when the need arises to combine several different theories to be able to understand a more comprehensive problem, their weak points are bound to be revealed by the inconsistency of the result, just as during the assembly of the separate parts of a machine one becomes aware of faults in their construction.

That so far embryologists have not yet succeeded in proposing a unitary view of development (from egg to adult, and encompassing all levels between external form and molecular machinery) should not be imputed so much to the lacunae that still exist in our knowledge as to the partial interpretations that have been proposed for the facts we do know. The analytical study of embryos began with the interpretation of aberrations in visible structure caused by microsurgical interference. The resulting concepts are still used as the basis of modern embryology, but to make them more precise and at the same time strip them of their philosophical overtones one must take into account the classical theories of biochemistry, genetics and cell biology as well. Thus, the science of development draws on all sources of biology, but in so doing may also absorb all sorts of erroneous ideas.

Because my own research was in an area where the discrepancy between observations and concepts was particularly flagrant, and because I found myself compelled to inculcate contradictory ideas on my students, I early on felt the need of an overall reconsideration of the problems of development, and finally decided to embark on the venture. For long years I have striven to give priority to the facts, to unearth forgotten work by others, constantly to question my own conclusions, and to seek a new language to express them in. As the broad outlines of the concept I have called Cell Sociology emerged, the transformation embryology has undergone and the dangers inherent in it prompted me to proceed with my synthesis. I see two such dangers. On the one hand the biochemical approach has developed to the detriment of studies at the level of visible structure, although the latter ought to be the censor of the former. On the other hand, theoretical approaches have undergone a sort of inflation and have attracted scientists from all fields – particularly physicists and mathematicians – who are the more inclined towards speculation as they lack works of a general nature to inform them about the facts and problems of embryology.

To be able to accomplish the task undertaken I needed not only encouragement but, more still, circumstantial criticism and complementary information. I have been fortunate in obtaining this cooperation from Dr. *J. Faber*, who for many years has been associated with the International Embryological Institute at Utrecht. His scientific erudition, his sympathy for new ideas, and his excellent knowledge of both French and English have made it possible for me first to publish the separate parts of the synthesis in a series of papers [*Chandebois* 1976b, 1977, 1980a, 1981]. When these partial studies led to a better understanding of the automation of development, the idea arose of explaining this insight in a book which would stand apart from the discussions published elsewhere and would be understandable to scientists who were not trained as embryologists. It was only natural that Dr. *Faber* should undertake the task of scrutinising and translating the French text, at the same time making many suggestions for improvement and adding additional information in places.

I owe a personal debt of gratitude to Prof. *M. Jeuken*, Editor of *Acta Biotheoretica*, for his interest in publishing my initial series of papers. Prof. *A. Wolsky* has done us the honour of showing much interest in the plans for the present book and we thank him for proposing to incorporate it into the *Monographs in Developmental Biology* series. We are the more grateful for this as the publication of unorthodox ideas usually meets with great difficulty. We thank the staff of the Library of the Hubrecht Laboratory for their bibliographical help and Ms. *Dorothy Parsons* for meticulously typing the various versions of the English text. Finally, we want to thank Dr. *D.A. Ede* for his kindness in writing the Preface and the publishers for their beautiful and efficient work.

Rosine Chandebois

Introduction

The development of an animal is governed by a programme contained in the egg, but because of the occurrence of metabolic processes the principles of its automation cannot be compared to those of man-made automata.

The development of a pluricellular animal encompasses a number of successive stages, each of which is characterised by a specific transitory organisation that is in general more complex than that of the preceding stage. By way of example figure 1 represents the development of the frog. We have chosen this example because the anurans, a subdivision of the amphibians, have always been the favourite experimental material of embryologists and, along with the urodeles, have furnished the most numerous and most important arguments for the discussion of general concepts.

From the instant the oocyte is woken from its lethargy by the entry of the sperm or by some other activating agent, development proceeds with the utmost precision. This is immediately apparent if we look at two kinds of document that every experimenter uses as the basis for his work: the so-called 'normal tables' and 'fate maps'. Normal tables, which have been established for many species, allow one to predict within narrow limits of accuracy at which times the various stages will appear if the time of fertilisation and the ambiant temperature are known. Fate maps are established by systematically placing marks of a vital stain in many accurately determined points on the surface of an early embryonic stage, and then noting their localisation on or within the embryo at later stages. Fate maps thus enable one to 'mark out' on the surface of the embryo, long before the actual organs appear, which group of cells will participate in the construction of an organ during normal, undisturbed development (this is called the 'prospective area' of the organ in question).

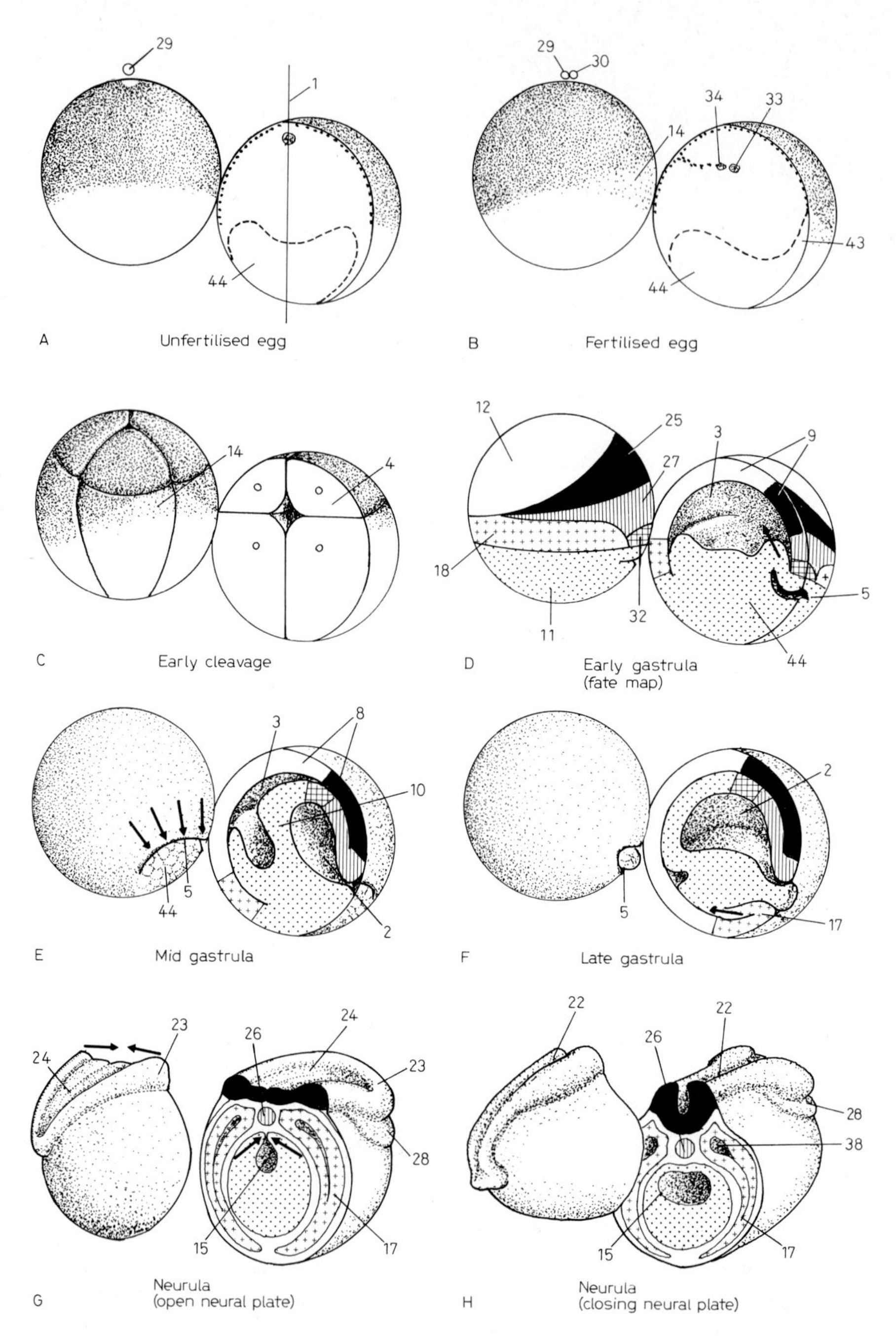

Fig. 1. The development of the frog. 1, animal-vegetal axis (A); 2, archenteron (E, F); 3, blastocoel (D, E); 4, blastomere = cleavage cell (C); 5, blastopore (D, E); 6, branchial clefts (I); 7, diencephalon (Ib, Ic); 8, ectoderm (E); 9, ectoderm (prospective) (D); 10, endoderm (E); 11,

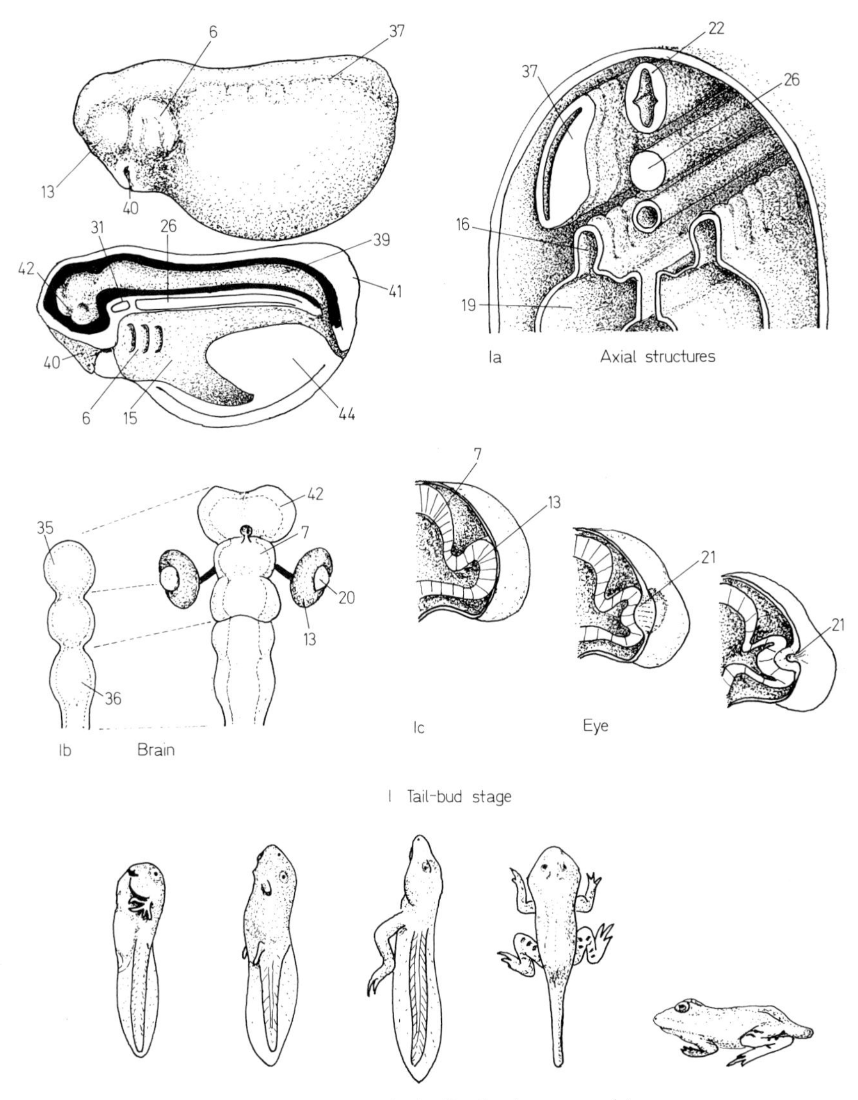

I Tail-bud stage

J Metamorphosis (functional organogenesis)

endoderm (prospective) (D); 12, epidermis (prospective) (D); 13, eye-cup (I, Ib, Ic); 14, grey crescent (B, C); 15, gut (G, H, I); 16, intermediary mesoderm (Ia); 17, latero-ventral mesoderm (F, G, H); 18, latero-ventral mesoderm (prospective) (D); 19, lateral plate (Ia); 20, lens (Ib); 21, lens placode (Ic); 22, neural axis (H, Ia); 23, neural crest (G); 24, neural plate (G); 25, neurectoderm (prospective) (D); 26, notochord (G, H, I, Ia); 27, notochord (prospective) (D); 28, placodal epidermis (G, H); 29, polar body (first) (A, B); 30, polar body (second) (B); 31, prechordal plate (I); 32, prechordal plate (prospective) (D); 33, female pronucleus (B); 34, male pronucleus (B); 35, prosencephalon (Ib); 36, rhombencephalon (Ib); 37, somites (I, Ia); 38, somites (prospective) (H); 39, spinal cord (I); 40, stomodeum (I); 41, tail bud (I); 42, telencephalon (I, Ib); 43, vitelline wall (B); 44, yolk (coarse) (A, B, D, E, I).

With the exception of the mammalian embryo, about which we will speak later, the only role of the environment is to ensure the survival of the individual during its development, particularly be enabling respiratory gas exchanges to occur. As a rule the environment is in no way involved in the emergence of the organisation of the embryo. At most, certain conditions in which embryos or larvae find themselves when they are in a critical stage can effect a choice between two (or more) patterns of organisation. For instance, in certain species the sexual characteristics of the gonads and genital ducts can be determined by the ambient temperature. In social animals the morphology and function of the adult animal depend on the composition of the society into which it is integrated.

The autonomy of the embryo and the precision of the transformations it undergoes show that development is an entirely automatic process: it represents the execution of a programme contained in the egg. The extreme complexity of the performance accomplished by this automaton can only be compared to the workings of a computer. This is the reason why in our days, with informatics revolutionising all areas of human activity, the science of development experiences a resurgence of interest and offers new fascinating perspectives for the understanding of life itself. For instance, it has been possible to simulate certain aspects of development on a computer, using programmes based on experimental data. Examples are: the growth of the normal and *talpid* chick wing bud [*Ede and Law,* 1969]; changes in the shape of the amphibian neural plate [*Jacobson and Gordon,* 1976]; mesoderm induction in the amphibians [*Weijer* et al., 1977]; cleavage of a molluscan egg [*Bezem and Raven,* 1975]. But the informatics revolution is far from having positive sides only. A number of theorists [*Arbib,* 1972; *Atlan,* 1972] have already proposed to apply automata theory to the most crucial problem of embryology: the progressive emergence of structure. The fault common to all these approaches is that the analogies that can be established between the genesis of an organism and the performances of man-made machines have been pushed too far. Strictly speaking the multitude of molecular species that make up the cytoplasm have nothing in common with the 'hardware' of a computer – if only because of the ongoing processes of metabolism. Where shall we find a machine that reproduces its own substance and at the same time grows, structures itself, reconstitutes itself upon trauma, reorganises itself in response to changes in its environment, and programmes its own reproduction? *The automatism of development has its own principles: we can only discover these by experimenting on living things.* The very fact that application of the same model of automatism to different aspects of development may lead to contradictory schemes of how things happen (a situation we will encounter already in the first chapter) shows that one should not use such analogies to explain phenomena for which experiment has not yet suggested an explanation.

For this same reason there is a real danger in simply transferring current and accepted terms from the language of information science to the terminology of embryology. When this is inevitable one must strip the words of certain connotations they have implicitly acquired by dint of being applied to automata. One must go back to rather broader definitions to make the terms apply to our knowledge of development as well as to the principles of automatism. This applies first of all to the word 'information' itself, which we will use in a broader (and therefore less exact) sense than the meaning it has in information science. Further, when we say that a part of an embryo contains the 'programme' for its development, what we mean is that its present properties commit it to a future concatenation of transformations that are rigorously determined independent of all external influences – we do not imply that the programme involves the use of a code. Similarly, our concept of 'cytoplasmic memory' should not give rise to the idea that we envisage a specific cellular compartment where molecules with a certain information content are stored – it is meant to express the fact that in an embryonic cell lineage a specific activity occurring at a given time leaves a trace of some sort in the cells, which makes possible other activities and more elaborate biosynthetic processes at a later time.

Consequently, the concept of developmental automatism that we will propose in this book takes very little directly from automata theory. Nevertheless, we can already say here that in a general sense this theory justifies the overall structure of the book, which otherwise could easily appear to run counter to common sense: we will analyse development before we analyse the egg, for the same reason that one must know how a computer functions before one can understand how the programmes are established that are fed into it.

The Automatism of Development

Cell Properties and Their Manifestation in the Features of Structural Patterns

Normal development consists of the exact reproduction of the structural patterns of the species; this involves the properties of the cell as a whole, such as cell growth, cell adhesion, cell death, etc. To understand the automation of pattern reproduction one must first understand the transformation and diversification of cell individualities.

In any animal species and at any developmental stage the form of the organism is virtually the same in all individuals, from the general shape of the body and its organs down to the most minute details, such as the decoration of the integument. Therefore ontogenesis must be considered as the reproduction of 'structural patterns' that are peculiar to the species. Ontogenesis implies an increasing complexity of organisation, which itself follows spatial and temporal guidelines rigorously set for every species; these are called 'spatio-temporal patterns'. Faithfulness of reproduction is an imperative condition for successful development. Deviations are generally disadvantageous to the organism if not incompatible with survival. Therefore, the ultimate aim of embryology must be to understand the automatism of the reproduction of structural patterns.

The forms of life exist by virtue of the conformation and arrangement of formed elements such as membranes, filaments and granules, which constitute the ultrastructure of living matter. These are 'structural complexes', i.e. assemblies of macromolecules (largely proteins). Everyone knows, and we will elaborate on this presently, that the conformation and arrangement of the formed elements result not only from the structure of their constituent molecules but also, and to a very large extent, from the physico-chemical conditions obtaining in the milieu that surrounds them. This is the 'ground cytoplasm', and more particularly the ions and products of intermediary metabolism contained in it. That is the reason why the form of a living being is considered to be the visible manifestation of its molecular activities. It is only one step from there to think that molecular biology is capable of solving – and will solve – all the problems of development. Many biologists have already taken this step, either by abstracting from form altogether or by establishing a direct relation between form and biochemical systems. However, such systems cannot maintain their activity outside the organisation of the cell. *The cell is the unit of life at the physiological level. At the level of morphogenesis it cannot be otherwise.*

The overall architecture of an animal, with each of its organs devoted to a particular function, reflects the division among the cells of the various tasks that are involved directly or indirectly in the animal's metabolism. This division of labour is today denoted by the term 'cellular differentiation'. To perform its specific task each cell type synthesises a specific kind of molecule: for instance, erythrocytes make haemoglobin, muscle fibres myosin, and the rods and cones of the retina rhodopsin. Because the most characteristic of these substances form part of structural complexes, the metabolic specialisation of a cell belonging to a functional organ – which today is called a 'fully differentiated cell' – is reflected in its form and ultrastructure or in the presence of specific organelles. For instance, a neuron is recognised by its dendrites and its axon; the cytoplasm of a muscle cell contains contractile fibres; and retinal cells carry a flagellum with vesicles at its base.

In fact the visible morphological features of the animal reflect not only the diversity in the ultrastructure of the cells as related to their functional specialisation. Rather, they reflect the diversity of the properties of the cells seen in their entirety. Among the most important of these properties is the adhesiveness of the cell surface [*Trinkaus,* 1967]. In the different tissues of the organism the cells do not adhere to each other equally strongly. They sometimes even repel each other. For instance, epithelial cells dispersed in culture regroup themselves, whereas cultured fibroblasts encountering each other stop moving and then move away from one another. Thanks to this 'differential adhesiveness' each tissue of the organism has its own particular texture, and when the cells are dispersed by appropriate treatment (e.g. removal of Ca, trypsinisation) they reaggregate and more or less re-establish the texture of the original tissue [*Moscona,* 1975]. Within one tissue adhesiveness sometimes varies from cell to cell. For instance, the same epithelium may take on different appearances (prismatic, cuboidal or flat-celled) according as the cells adhere to each other very strongly (with almost their whole surface) or less strongly. In addition, cell adhesion is selective in that a given cell type behaves differently when confronted with cells belonging to various other cell types of the same organism. For instance, if prospective ectoderm and endoderm from a blastula are placed side by side in saline they repel each other; if a piece of prospective mesoderm is placed in between, it adheres to the two others [*Townes and Holtfreter,* 1955]. Embryonic cells do not exhibit this diversity of adult cells. During the first stages of development they all resemble each other and all have an equally simple appearance: one says that they are 'undifferentiated'. It is only gradually, while cell morphologies diversify, that the features of the adult organisation become delineated and refined. Hand in hand with this goes a progressive transformation and diversification of the cells' biochemical activities; transformation and diversification are two aspects of what we call differentiation, taking the term in its dynamic sense.

It should be pointed out here that embryologists mean something else by the term 'cell differentiation' than cytologists and molecular biologists. The term differentiation embodies a heterogeneous complex of elements. The adjective 'differentiated' denotes the complex, not the elements. The term 'differentiation' in its static sense denotes the *diversity* of the elements, but in its dynamic sense it also denotes their *diversification,* that is, their becoming more and more different from one another. The term was first introduced to denote the diversity of cells in an organism as it is observed histologically (comprising both the diversity of tissues and the diversity of cell morphologies within one and the same tissue). It was retained in the same sense by embryologists (who were looking at the cells as they are integrated into an organism), but they also extended it to the diversification of cell types during ontogenesis. Cytologists primarily study the morphology of the various organelles in the cells as such, independent of the organism, and therefore introduced the terms 'differentiated', 'undifferentiated' and 'highly differentiated cell'. However, they have also used the term 'differentiation' in its dynamic sense, meaning in this case the increasing complexity of the cells' ultrastructure in relation to the establishment of their definitive functions. Various authors have already pointed out that in the study of development the adjectives 'differentiated' and 'undifferentiated' as applied to cells are incorrect and confusing. However, they are so much hallowed by long use that there are no equivalents for them and they cannot be avoided [*Chandebois,* 1976a]. Finally, molecular biologists have restricted the meaning of the term still more by their usage. They use it to denote the state of cells possessing the morphological and biochemical characteristics of a given tissue, to the exclusion of all progressive transformations that lead up to that state during embryogenesis.

Visible organisation is not determined instantaneously. During the progression of differentiation certain properties that cells possess at a given time leave behind indelible traces which, together with the traces of earlier activities, lead to more and more complex structures. This is particularly the case with respect to changes in cell adhesion. In some areas the cells establish more intimate contacts, while in other areas they break their contacts, after which some may move until they make contact with another cell type. From this there results either mass movement of cells or migration of individual cells, the direction and amplitude of which rigorously depend on the properties of both the moving cells and the cells they encounter. All this leads to an increase in complexity that is often dramatic (e.g. invagination of the archenteron, closure of the neural plate, segregation of mesodermal derivatives). The spatial and temporal precision of these cellular displacements reminds one of the art of choreography, hence the evocative term 'morphochoresis' (from French: 'morphochorèse').

Growth also plays an important role in the reproduction of structural patterns. It usually takes the form of cellular multiplication (hyperplasia). In exceptional cases cells stop dividing but grow in size (hypertrophy), as in the case of neurons in the vertebrates. In addition, within one tissue the cells do not always divide synchronously. Therefore, not all body parts and organs expand proportionately. This phenomenon is called 'differential' or 'allometric' growth [*Huxley,* 1932] and plays a paramount role in the shaping of certain organs. For example, distinct regional differences in the growth of the walls of the neural tube lead to the formation of the five primary vesicles of the brain, each with its particular shape. These features persist into adult life, whether the neurons retain or lose their capacity to divide. Finally, cell death is crucially involved in the shaping of most organs [*Saunders and Fallon,* 1967]. In certain areas of the embryo cells undergo a process of self-destruction, often in large numbers, at a precisely timed stage of development. For instance, the separation of the digits in the hand and foot plates of higher vertebrates results from the death of cells lying between the skeletal rays; in cases where it does not occur an interdigital membrane develops, as in the webbed feet of the duck.

Visible structure does not exhaust the complexity of the animal's organisation. Certain of its features are invisible even in the electron microscope. Only experiments can bring them to light. A vital dye taken up by a tissue under conditions of anaerobiosis loses its colour because of reduction by respiratory enzymes. If groups of small organisms or embryos of the same species and age are thus treated, the dye is progressively discoloured in the same manner in all individuals. Discolouration may for instance start at a given point in the body and from there gradually extend all over it. The existence of such invisible features is fundamental to the interpretation of experimental results and originates from the fact that synthesis and degradation do not proceed at the same rate in all cells of the organism: the cells do not have the same 'metabolic standing' [*Chandebois,* 1976a]. One can make a practical distinction between 'overt' and 'covert' patterns, but in fact we have to do with the same structural pattern, produced by one and the same cell population, which has both visible and invisible features – much as a bank note carries both a directly visible printed design and a watermark that is only discernible under certain conditions of illumination [*Chandebois,* 1977].

In conclusion, the existence of visible and invisible structures cannot be explained directly from the properties of the constituent molecules. It originates from the fact that each cell possesses a particular 'individuality' (or 'personality'), which distinguishes it from other cells and comprises both visible and invisible features. This individuality is defined by the peculiarities of its qualitative (molecular composition) and quantitative metabolism (metabolic standing) and by their manifestation in the various cell properties (growth, adhesiveness, cell death, etc.). The fact that the organisation of the individual organism conforms to the structural patterns of the species implies that all aspects of the individuality of any cell depend on its position in the whole. This is what one calls 'positional effect'. The reproduction of structural patterns is effected through the progressive transformation and diversification of cell individualities. In consequence, at any time during an animal's life *it is to the genealogy of each cell type that we must look to find out how the animal's organisation is determined.* To simplify the study of the automatism of the reproduction of structural patterns we will therefore first establish the principles of the dynamics of differentiation. However, before approaching the problem from that angle we must first answer the following basic question: Why is it that a cell at a given moment in time possesses an individuality that distinguishes it from other cells and depends on the cell's position among the other cells?

The Individuality of a Cell in the Organism, and How It Is Determined

The structural complexes whose conformation determines the ultrastructure of all cells, almost without exception contain structural proteins associated with non-protein molecules such as lipids, carbohydrates, etc. Each protein itself is an assembly of polypeptides linked together by non-covalent bonds.

Each polypeptide is a chain of amino acids linked together by peptide bonds. The number and identity of those amino acids, as well as the order in which they are arranged, define the primary structure of the polypeptide. Once its synthesis is completed the polypeptide chain folds in a certain manner because, under the influence of the physico-chemical conditions of the molecular environment, certain radicals of the chain become linked by non-covalent bonds. The chain then assumes a tertiary structure that depends not only on the position of its radicals – i.e. on its primary structure – but also on the physico-chemical conditions obtaining in the ground cytoplasm. This contains the products of intermediary metabolism, the nature of which depends on the properties of enzymes. We may say, therefore, that the tertiary structure of structural proteins depends indirectly on the primary structure of the enzymic polypeptides; this also determines the conformation of the non-protein molecules that take part in the construction of the structural complexes. If we disregard all external influences, we can therefore say that the morphology of a cell is defined by the primary structure of the various types of polypeptide it produces.

The morphology of a cell can be remarkably constant notwithstanding continuous molecular renewal and cell division. This implies rigorous control over the sequential order of amino acids during the synthesis of polypeptides. Every polypeptide chain is constructed in the cytoplasm of the cell on a template: the messenger RNA (mRNA), which is a chain of nucleotides. With the help of two other species of RNA (transfer RNA and ribosomal RNA) the amino acids arrange themselves in a sequence that corresponds to that of the nucleotides in the mRNA molecule. That is why one calls the synthesis of a polypeptide 'translation'. The mRNA molecule itself is a portion of a molecule that was originally longer (called nuclear or heterogeneous RNA, HnRNA), from which it is excised by a ribonuclease. mRNA is of itself very short-lived but its lifetime can be considerably enhanced by 'masking' or stabilisation (by certain proteins) prior to translation; after its translation it is degraded, and it must therefore be continually remade by the cell. HnRNA is constructed on a metabolically stable template: a segment of a long molecule of DNA called the 'operon'. Its synthesis (called transcription) starts at one end of the operon with the help of an enzyme called RNA polymerase. Its nucleotides are arranged in a sequence corresponding to that of the nucleotides in the DNA. The mRNA portion is the last part of the molecule to be synthesised and corresponds to a particular transcription site in the operon: the 'structural gene'. The first and longest part of the HnRNA molecule, called Hn1RNA, does not leave the nucleus and is not translated. It contains repetitive nucleotide sequences, whereas the nucleotides in mRNA differ from one gene to the other [*Georgiev,* 1972].

Recently it has become clear that the notion of structural gene is more difficult to define than was originally assumed. Firstly, one has found that some mRNAs are mosaics representing an assembly of sequences transcribed from different chromosomes [*Darnell,* 1978]. Second, the initiation and termination sites on the operon seem not to be rigorously fixed [*Gilbert,* 1976].

Because the primary structure of the polypeptides is encoded in the structural genes, the morphology of a cell is usually considered to be the transposition in three-dimensional space of the linear sequence of the nucleotides in the DNA. This is expressed by saying that the DNA contains the 'genetic information'. Through the replication of the DNA and the equal distribution of the chromosomes during cell division the genetic information is transmitted to the cells from one generation to the next. Nevertheless, the important role of the DNA should not be allowed to obscure that of the ground cytoplasm. Cells are not closed systems. As they take up molecules from outside, the physico-chemical conditions in the cytoplasm are modified. These modifications not only affect transcription and DNA replication [*Beljanski,* 1983] but also the biosynthetic steps following translation during the construction of structural complexes [*Koshland and Kirtley,* 1967].

One usually restricts the discussion of the problems of differentiation to those of the production of the proteins that enable the tissues to perform their functions in the organism. However, for the study of morphogenesis it is just as important to enquire why the cells of one and the same tissue differ in appearance and properties, indeed often extremely so.

Histotypic Protein Synthesis

The cells of a tissue exhibit not only a dominant 'luxury' metabolic strategy (which enables the tissue to exert its specific function) but also several latent luxury metabolic strategies, which are dominant in other tissues. As a result of a change in cell contact relations one of the latter may become dominant (transdifferentiation). Once a metabolic strategy is extinguished the cells are unable to re-establish it. This suggests that the transcription of a structural gene is accompanied by the synthesis of a molecule (probably an RNA 'primer') that can specifically initiate the further transcription of that same gene.

The production of specific substances (notably proteins) which enable a tissue to perform the function assigned to it in the organism is reflected in certain features of the ultrastructure of its cells. If the tissue is manipulated so that it ceases to function in this manner, these features fade away. The cells reacquire a generalised, deceptively 'simple' appearance: they are said to dedifferentiate. At the same time they undergo what is called 'cellular activation'. Replication and transcription are intensified: mitotic activity increases, voluminous nucleoli appear in the interphase nuclei, and the ground cytoplasm is invaded by ribosomes [*Hay,* 1968]. Obviously the proteins which enable a tissue to perform its specific function are not indispensable to the survival of its cells. That is why one calls them 'luxury proteins' (as against 'household proteins'). The presence of those proteins (whose activity is often bound up with the presence of non-protein molecules also produced by the cells) cannot be dissociated from specific chains of biosynthesis and degradation. It therefore seems preferable to say that in order to carry out its specific function a tissue must maintain a specific 'luxury metabolic strategy' [*Chandebois,* 1981]. However, this should not be considered to be the exclusive property of one tissue. The specific proteins or mRNAs (or both) involved in it may sometimes be found in other tissues as well, albeit in much smaller quantities [*Clayton,* 1982]. For instance, the mRNA for ovalbumin, a protein specific for the chick oviduct, is also produced by the spleen, the heart, the brain and the liver [*Tsai* et al., 1979]; globin RNA sequences have been detected in the nuclei and cytoplasm of non-erythroid tissues such as brain, liver and cultured fibroblasts [*Humphries* et al., 1976]; glial cells contain both actin and myosin [*Orkin* et al., 1973]. Actin, which is considered specific for muscle, occurs also in blood platelets, fibroblasts, spermatozoa and nerve cells [*Pollard and Weihing,* 1974]. (In this case it is possible that the protein concerned also has a functional role in those cells, because actin is known to be part of the 'cytomusculature' of many cells and to be involved in changes in cell shape.) As we shall see later, an even broader spectrum of specific proteins has been observed in embryonic cells and tissues. From all these results one may conclude that one tissue maintains several 'luxury metabolic strategies' simultaneously. One of them is 'dominant': this is the one that ensures the tissue's particular function and specific histological appearance. The other strategies are 'latent' and without apparent significance for the activity and structure of the cells, although, as we shall see below, not without significance for its behaviour in experimental situations.

We may conclude that tissue differentiation is based essentially on *quantitative* differences in the synthesis of specific substances. It should be emphasised that this also holds for the synthesis of the household proteins. A classical example is that of the lactic dehydrogenases, a family of five enzyme forms or 'isozymes', each of them consisting of four polypeptides. The latter belong to two different classes (A and B) and associate at random. The spectrum of enzyme forms found in a tissue depends on the relative amounts of the two polypeptide classes and varies considerably from one tissue to another [*Markert and Møller,* 1959].

The fact that not all cells of an organism produce the same proteins of course means that not all genes are actively transcribed in all cells. What we may infer is a 'differential release of genetic information'. Direct proof of this is furnished by the giant chromosomes in the salivary glands (and some other organs) of dipteran insects. These chromosomes contain multiple copies of despiralised chromonemata, filaments of DNA which each show a large number of clews situated at irregular intervals, called chromomeres. The chromomeres are arranged in register in the chromosome, giving the appearance of alternating bands and interbands. At certain sites in the chromosome the chromomeres are unwound and the loops they form together give rise to a so-called 'puff'. Messenger RNA synthesis can only be demonstrated in these puffs [*Beermann,* 1966], which consequently represent active genes while the bands represent repressed genes. Gene activity is regulated by substances entering the cell from outside or present in the cytoplasm. When the salivary gland is exposed to ecdysone (the moulting and metamorphosis hormone) certain new puffs appear in a rigorously determined temporal sequence [*Clever,* 1961]. When giant chromosomes from the salivary gland are transferred by micromanipulation to the cytoplasm of eggs of the same species at different cleavage stages, the existing puffs are resorbed and transform into bands, while new puffs appear at specific sites differing according to the stage of the egg [*Kroeger,* 1960].

The fact that an enucleated cell generally does not survive suggested long ago that the DNA is at the controlling levers of the cell. Hence the habit of only considering the activity of the genes when discussing the problems of differentiation. However, this idea ought no longer to be maintained. Rather, the nucleus is comparable to a stencilling workshop, each gene being a 'stencil' from which multiple copies are made. But the workshop operates on the 'self-service' principle: it only works when there are 'customers' (the cytoplasm) who cause copies to be made from certain stencils.

The regulation of transcription, the understanding of which is of paramount importance for the study of development, is far from being completely elucidated for the case of eukaryotic DNA. Nevertheless, differentiated cells, when experimented upon in vivo or in culture, show certain properties which allow us to divine certain characteristics of gene regulation. We must know these properties before attempting to make a choice from the various models of gene regulation proposed by the biochemists.

Generally, when tissue cells are cultured in vitro their dominant luxury metabolic strategy remains unchanged, even when it temporarily stops manifesting itself during a period of intense proliferation. Muscle fibres retain the capacity to contract, glandular cells the capacity to secrete the same products. The differentiated characteristics of the cells appear to be stable and transmissible after the fashion of hereditary traits. It was *Gurdon* [1973] who already drew the inescapable conclusion from this: a structural gene seems to be capable of maintaining its own activity indefinitely. This suggests that the transcription of a gene is accompanied by the production of a molecule that can *specifically* re-initiate ('prime') the transcription of that same gene. *Gurdon* [1973] retained the idea proposed earlier by several biochemists that this molecule is a species of RNA.

Nowadays one thinks that perhaps the part of the HnRNA that is not translated (Hn1RNA), and is synthesised before the messenger, is involved in priming the transcription of the corresponding gene. In fact, short RNA molecules called chromosomal RNA (cRNA) are often found associated with the chromosomes. Their nucleotide sequences vary from tissue to tissue. In regenerating liver the appearance of a new protein is preceded by that of a new type of cRNA, which has repetitive sequences in common with the new type of HnRNA produced [*Mayfield and Bonner,* 1971, 1972]. Further, short

RNAs extracted from nuclei of a given tissue are capable of stimulating the transcription of RNA from chromatin of the same tissue [*Kanehisa* et al., 1974].

Thus, cRNA meets all the requirements for a molecule that can specifically activate the genes involved in the various tissue functions. Very different models have been proposed for the manner in which Hn1RNA could reactivate the genes [*Britten and Davidson,* 1969; *Frenster and Herstein,* 1973; *Holmes and Bonner,* 1973]. More recent work has shown that transcription in vitro proceeds without a primer [*Anthony* et al., 1966], while a newly synthesised DNA molecule is always covalently bound to a short RNA [*Gefter,* 1975]. Hence it is possible that Hn1RNA is indirectly involved in the priming of transcription if the latter is preceded by the replication of the gene, as suggested by the work of *Bell* [1971]. Obviously the hypothesis of gene regulation by short RNAs is still unproven and not generally accepted. If we retain it here it is because it is at present the one that agrees best with all we know about development, as we will show repeatedly in the rest of this book.

Although differentiated tissues generally show a certain stability, they may sometimes change their dominant metabolic strategy, either in vitro or as a result of experimental manipulation. Today this is known under the name of 'transdifferentiation'. We will mention two classical examples. In adult newts and in embryos of other vertebrates the iris of the eye or the optic vesicle is capable of giving rise to a so-called Wolffian lens regenerate when the original lens, which is of epidermal origin, is removed. When a leg of a urodele amphibian is amputated the cartilage cells of the stump are capable of taking part in the regeneration of muscle tissue, and conversely the muscle cells may form cartilage [*Trampusch and Harrebomée,* 1965; *Wallace,* 1981]. Before a tissue can abandon its dominant metabolic strategy certain specific cytoplasmic components must be shed from the cells (pigment granules in the case of the iris) or must be broken down (muscle fibres in the case of muscle). The establishment of a new dominant metabolic strategy reflects a histogenetic potency, a certain predisposition of the tissue. In the case of the optic vesicle of the chick embryo the lens potency must be bound up with a subliminal synthesis of crystallin, the lens-specific protein. Crystallin and the corresponding mRNA have indeed been detected in the optic vesicle between the 3rd and the 8th day of incubation, which is exactly the period during which it is capable of forming a lens in vivo [*Clayton* et al., 1979]. Their production is considerably enhanced during transdifferentiation [*Thomson* et al., 1979]. It is clear, therefore, that *transdifferentiation means that a hitherto latent metabolic strategy supersedes the strategy that was dominant until that time.*

Further, the disappearance of both crystallin and its mRNA coincides with the definitive loss of lens-forming potency. This suggests that *the disappearance of the products of a particular gene definitively puts an end to the possibility of this gene being reactivated further.* This would tally very well with the concept of the regulation of transcription that we favour, because the synthesis of the putative specific 'primer' of a gene, the Hn1RNA, is not dissociable from that of its messenger (and the corresponding protein).

Transdifferentiation is not caused by some substance produced by another tissue which would, for instance, directly activate the genes coding for new specific substances; it probably results from a change in cell contact relations. When the cells of the optic vesicle of the chick embryo [*Eguchi and Okada,* 1973; *Okada,* 1975; *Clayton* et al., 1977; *Okada* et al., 1979] or those of the newt iris [*Eguchi* et al., 1974] are cultured in vitro, some of them aggregate locally into so-called lentoids and take on the appearance of lens fibres. But not all cells change. Certain culture conditions (amount of medium, bicarbonate concentration) favour transdifferentiation but do not evoke it [*Pritchard* et al., 1978]. It is probable that the maintenance of the differentiated state of a tissue depends on its coherence, which in vivo is maintained by the influence of neighbouring tissues. If as a result of experimental manipulation the cell contacts are loosened, this apparently leads to elimination of specific components of the cytoplasm, which evokes the dedifferentiation that makes transdifferentiation possible [*Lopashov,* 1977]. The role of cell contact relations in the maintenance of the dominant metabolic strategy is well illustrated by experiments made on the iris of the adult newt [*Eguchi and Watanabe,* 1973]. Its ventral portion does not normally transdifferentiate in culture, but it can form lentoids after treatment with a carcinogen that reduces cellular adhesiveness. It seems, therefore, that *the relative intensities of luxury metabolism in a tissue depend on the manner in which the cells are associated with each other; this in its turn depends partly on mutual cell adhesiveness and partly on the physical and chemical properties of their environment.* In other words, the rate of gene transcription would be controlled to a large extent by signals received by the cell periphery [*Moscona,* 1975]. This idea can no longer be doubted and is supported by some findings of molecular biologists. In cultured fibroblasts mRNA synthesis and DNA replication are profoundly affected by the shape of the cells [*Ben-Ze'ev* et al., 1980], which no doubt affects the state of the cell surface. In cultured cells the cell membrane influences DNA replication [*MacDonald* et al., 1972]. Obviously these aspects of gene regulation, which are essential for the understanding of development, are so far entirely unexplained by molecular biology.

There are probably also controls at the level of the translation of genetic information. Among other things, various authors have found indications that short-lived mRNA transcribed during a period of proliferation in vitro is translated preferentially to stabilised mRNAs coding for luxury proteins [*Whittaker,* 1965]. In addition one often sees that whole chromosomes or considerable amounts of chromatin are eliminated from the nuclei of somatic cells in early embryonic stages, which irremediably restricts the possibilities of differentiation of the cells in question.

The Diversity of Cell Individualities in the Various Tissues

> The genetic information set free by the differential transcription of the DNA is filtered and accommodated through various factors intervening during the phases that succeed transcription – particularly in the establishment of non-covalent bonds. These factors are to a large extent substances which enter the cell from the intercellular milieu, such as ions, substrates, vitamins and hormones. Their diversity partially explains the variety of cell individualities encountered within one and the same tissue.

Of the factors that determine the extreme diversity of cell individualities in the various tissues we know even less than of those involved in tissue-specific protein synthesis. This is because they probably act at all levels of the many different chains of biosynthesis and degradation.

What we for convenience call a luxury protein specific for a particular tissue in reality often consists of various families of proteins with slightly different electrophoretic properties. Their numbers make it unlikely that each of them is controlled by a specific gene but can probably be explained by assuming that the mRNA associated with the polyribosomes (where it is translated) does not correspond to a strictly delimited transcription site on the DNA, but that the initiation and termination sites can be displaced as a function of the physico-chemical conditions obtaining in the extrachromosomal milieu [*Beljanski,* 1983]. As a result the relative abundance of various families may vary from one cell type to another in one and the same tissue. A good example is crystallin in the chicken. There are three families of α-,

β- and δ-crystallin, each consisting of various kinds of molecules. The lens fibres contain more δ-crystallins than the cells of the cortex, where the α-crystallins predominate [*Genis-Galvez* et al., 1968].

Even when a cell has adopted a certain repertory of polypeptides this does not mean that the nature of its structural complexes is fixed for good [*Koshland and Kirtley,* 1967]. Their construction is greatly subject to chance, as shown by the alteration of certain specific cell activities that may ensue from a simple change in pH. A given type of non-covalent bond can only be established under specific physico-chemical conditions. The composition of the ground cytoplasm must be appropriate for the sort of structural complex to be constructed, otherwise polypeptide chains will fail to associate or take on a faulty tertiary structure. Non-protein molecules are also indispensable for the construction of a structural complex. Often the extracellular environment must supply these or furnish the substrates needed for their synthesis. Enzymic proteins must assume an adequate conformation to be able to attach themselves to the substrate molecules [*Ebert and Kaighn,* 1966]; if they fail to do so they will remain inactive. In many cases enzymes function only when associated with a co-enzyme, which most of the cells or even the organism as a whole cannot synthesise. They are generally taken up with the food (vitamins). If for some reason they fail to reach the tissue this may lead to a drastic change in the cell individualities. This is exemplified by experiments performed with the oesophageal epithelium of the chick embryo [*Fell and Mellanby,* 1953]. Normally this tissue desquamates while producing keratin. If it is cultured in the presence of vitamin A it forms a mucous epithelium that is sometimes ciliated. If vitamin A is then excluded from the culture medium the tissue resumes its normal configuration. It should also be noted that the manner in which protein and non-protein molecules assemble into a structural complex often represents only one out of many possibilities. Why the molecules do not assemble at random cannot be said at present.

At any rate we can state that a process of 'filtering' of the genetic information continues after it has been translated on the principle of the differential release of genetic information (p. 15). At all subsequent biosynthetic steps whole molecules or parts of molecules are eliminated or inactivated, or serve no purpose. In addition, the information that has thus been filtered is then 'accommodated' by the cell in a certain manner. On the one hand certain biosynthetic chains take precedence over others, while on the other hand gene products are utilised in various different manners according to the nature of the non-protein molecules that are synthesised within or procured from outside the cell.

We know next to nothing about the factors that are responsible for the infinite variety in metabolic standing and adhesiveness of cells. Obviously the cytoplasmic composition plays a role, particularly the presence of histotypically specific substances, for tissues differ both in structure and in proliferation capacity. For instance, liver cells continue to be able to divide in the adult, whereas in higher vertebrates nerve cells lose this capacity early in development. On the other hand, it has been possible to show experimentally that the cellular environment also plays a role. Epidermal cells can greatly increase their mitotic activity and assume the appearance of an infiltrating tumour if the underlying dermis is suitably manipulated [*Seilern-Aspang and Kratochwil,* 1965]. The cyto-architecture of an epithelium (prismatic, cuboidal or flat-celled) depends on the strength of mutual cell adhesion; it can be modified in vitro by association with various kinds of connective tissue [*MacLaughlin,* 1963]. Thus, the metabolic standing and the adhesiveness of cells, just as the other features of cell individuality, are determined both by the extrachromosomal milieu and by the extracellular environment.

Summarising, the transcription of certain genes specific for a tissue, which seems largely to be controlled by cell contact relations, provides to the cells a 'theme' on which numerous quantitative and qualitative variations are still possible, depending on a multitude of factors, among them diverse substances coming from outside the cell.

The Functioning of the Cell in the Organism in Terms of Automatism

The process by which a cell filters and accommodates the genetic information contained in its DNA implies that at any moment in time a cell registers extracellular information – particularly positional information provided by neighbouring cells. It interprets this information as a function of the cytoplasmic information it had stored previously, which depends on its previous activities. If one only considers the way in which a cell's individuality is determined (not its multiplication), the cytoplasm may be compared to the 'memory' of a computer. This stores the 'input data' as well as the 'results' of the computations performed. These together establish a 'programme', which is submitted to the DNA. The latter therefore seems to be comparable to the 'arithmetical and control circuits' of a computer (part of the 'hardware').

The considerable gaps that still exist in our knowledge of the molecular machinery of differentiated cells fortunately do not prevent us from sketching a broad outline of the automatism of development. We have already pointed out that the production of a complete repertory of substances is not a sufficient condition for a viable organism. The structural patterns of the species must be correctly reproduced. Understanding development essentially means understanding the effect of position. Therefore, the nature and the target of the factors that ensure the filtration and accommodation of the genetic information are only of secondary interest compared with the problem of the *origin* of these factors.

It is a well-established fact that the individuality of a cell changes when a modification takes place in its environment, cell growth and cell adhesion being more susceptible to change than the specific activities of the tissue. In the organism the environment of a cell is essentially represented by other cells with which it is in contact or from which it is separated by a narrower or wider space occupied by a cellular matrix. The cell communicates with its neighbours not only through specific or less specific substances that it secretes into the intercellular space, but also through other signals whose nature we usually do not understand, and for which the cell periphery serves as a receptor. In this way *at any moment in time every cell registers information, which we call 'positional information' because it determines features of the cell that depend on its position in the organism* [*Chandebois,* 1976a, b]. It should be stressed that we do not use this term in its original sense as defined by *Wolpert* [1969]. In fact, the idea that a cell is informed of its position by some diffusing substance or by some other signal emanating from a remote region has never been convincingly shown to be true. We will therefore assume that positional information is only communicated to a cell by other cells that are topographically adjacent to it. In addition, various other substances are dispersed by way of the intercellular spaces: substrates and vitamins, as well as, in those animals that possess a circulation, hormones secreted by the endocrine glands. All these substances, widely distributed throughout the organism, constitute complementary extracellular information which is superimposed on or modifies the positional information. Experience shows that extracellular information does not only act at the level of the filtration and accommodation of genetic information becoming available at a given time: cells isolated in culture do not lose all the features of their individuality, and the same factor acts differently on different cells. Consequently *differentiated cells obviously also use a form of stored information, the source of which must be their previous activities (and those of their ancestors). This information we will call 'cytoplas-*

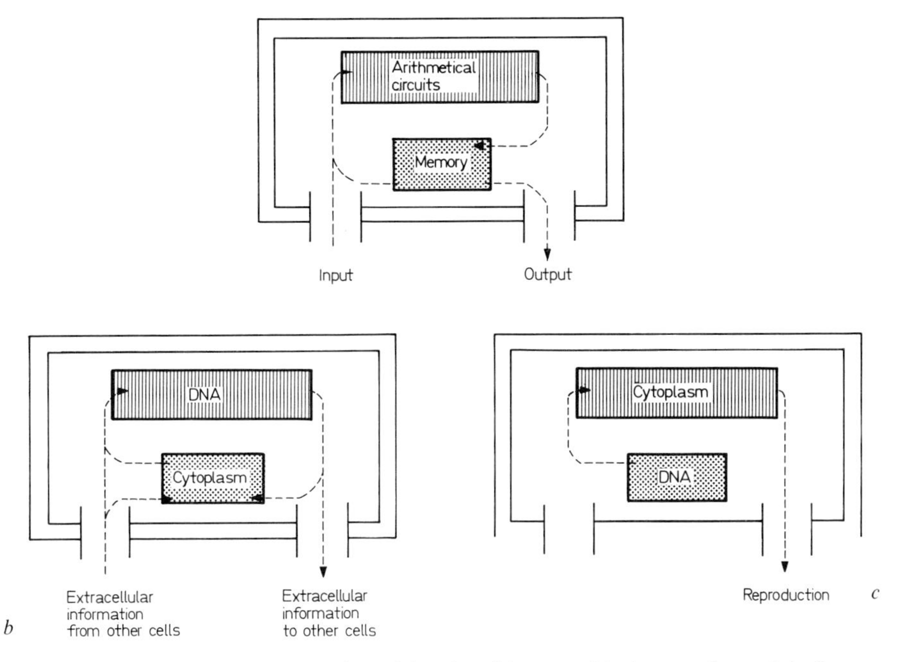

Fig. 2. Comparison between the activity of a cell integrated in the organism and the functioning of a computer. *a* General scheme of a computer. *b* Possible scheme of the automatism of the specific activities of a cell. *c* Possible scheme of the automatism of the reproduction of a cell.

mic information' [*Chandebois,* 1976a, b], a term which should be taken also to encompass those characteristics of the plasma membrane which influence the activities of the cytoplasm and the nucleus. *Two cells can only have exactly the same individuality if they receive the same extracellular information at the same time, provided their ancestors, at least in the recent past, have gone through the same activities.*

The functioning of a cell in the organism, depending as it does in every particular on the information the cell has previously registered, shows a striking analogy to the workings of a computer (fig. 2). In order to make the comparison as precise as possible, we shall start with the concise definition given by *MacCarthy* [1966]: 'A computer, as hardware, consists of input and output devices, arithmetical and control circuits and a memory. Equally essential to the complete portrait is the programme of instructions, the "software" that puts the system to work. The computer accepts information from its environment through its input devices; it combines this information, according to the rules of the programme stored in its memory, and it sends information back to its environment through its output devices.' The main activity of any cell of the organism is to synthesise specific substances, a part of which is exported from the cell and, for this reason, resembles the output of a computer. Extracellular information, which is registered continuously, is analogous to the input of the computer. This information is combined with previous information (particularly part of the previous 'output') which has been stored in the cytoplasm. The latter can be compared to the memory of the computer. This analogy is supported by the fact that in many instances cells can dedifferentiate and then redifferentiate in another manner. In such cases the cytoplasm is like a 'random access memory' (i.e. a memory which can be read, erased and rewritten). When cells retain their individualities whatever their environment, the cytoplasm is like a 'random output memory' (i.e. a memory which can be read but neither erased nor rewritten). In the differentiated cell viewed as a computer, a certain analogy exists between the DNA and the arithmetical circuits. This view is reinforced by the fact that during development cells divide and progressively change their individualities. It is evident that all kinds of molecules whose nature is changed belong to the 'software', i.e. specific substances released (output), substances present in the cytoplasm (information stored in the memory) and substances which have entered the cell from outside (input). On the other hand, the 'hardware' must be exactly reproduced. Consequently the DNA obviously belongs to the 'hardware'. It is the most important functional unit, without which no computation is possible. It has a peculiar, species-specific design, so that it determines the *specificity* of the molecules to be synthesised which constitute the output (cf. the metaphor of the 'stencilling workshop', p. 16).

This scheme of the functioning of the cell superficially seems to be at variance with the one that is generally accepted today, but this is by no means so. The model that has been used in the application of information theory to cell activities is the bacterium. This cell type is unable to build up a differentiated organism, its 'output' being molecular renewal and self-reproduction only, processes for which genetic information stored in the DNA is required and sufficient. Here the DNA is like a 'memory' peculiar to each species, from which information is selected in order to establish the programme of cell reproduction. This idea seems the more appropriate since DNA surprisingly resembles the tapes on which programmes are encoded (for the same reason, DNA has also been compared to a coded message in a telecommunication machine [*Atlan*, 1972]). However, the classical scheme obviously refers to a *different* aspect of cellular automatism: the reproduction of the 'hardware'. In pluricellular organisms the importance of cell division is superseded by that of differentiation, so that in this case we must altogether discard the classical scheme to prevent confusion.

A cell integrated in the organism cannot be considered in isolation. It continually registers information communicated to it by its neighbours (or by remote cells in the case of circulating substances), interpreting it in terms of its own previous activities. Some theoretical biologists view the organism as a functionally independent 'input-output system' [*Rosen,* 1972]. However, in a cell population the units multiply, diversify, renew themselves, 'emigrate' or destroy themselves singly or in groups, exchange information, and divide collective tasks among themselves, thus achieving a functional equilibrium. A cell population is therefore much more akin to a human society than to a network of automata. A cell in isolation can neither maintain its activities unchanged nor respond to a stimulus that could transform *a group of cells of the same type.* In other words, cell individuality is not based on individual memory alone: we have to do, to different degrees in different tissues, with a 'group effect' implying a 'collective memory'. And it is this aspect of the social behaviour of cells that underlies the phenomenon of progression that we call development. Its course is in many respects comparable to the history of a civilisation, and its study must be pursued from the viewpoint of a sociology of cells [*Chandebois,* 1976a, b]. The difference is that this 'civilisation' time and again reproduces itself unchanged in the individual organisms of a species and unfolds in an entirely automatic manner, because its members are 'automata' [*Chandebois,* 1980a]. To each member is assigned a specific individual programme which enables it to *contribute to the execution of the communal programme of the whole.*

The Dynamics of Differentiation

As we have already said, during development the organs of the animal gradually appear. In parallel with this, the various cell populations in a stepwise process acquire the configurations typical of the various adult tissues. Subsequently the populations become further structured as a result of the extreme diversification of cell individualities in each tissue. This process of differentiation proceeds by way of successive subdivisions. In other words, there where at a given early stage we find a region in the embryo that looks homogeneous, at a later stage we will find various distinct regions, which will then subdivide further. For instance, in the amphibians (fig. 1) the surface of the gastrula is uniform in appearance. At the neurula stage the neural plate emerges. After having folded in on itself to form the neural tube, at the tailbud stage it will show the rudiment of the brain, which only consists of three primary brain vesicles instead of the later five. When the optic vesicles emerge from the forebrain they are not yet segregated into neural retina and pigment epithelium. Similarly, the mesodermal mantle of the gastrula is still uniform, but in the advanced neurula it has given rise to various structures, some of which will still subdivide further: the notochord, the rows of somites, the intermediate mesoderm, and the two lateral plates. Therefore, in the course of normal development the descendancy of a given part of the embryo at a later stage or in the adult comprises various types of differentiated cells, which together represent the normal products (or 'achievements') of the part in question at that stage. For instance, in the vertebrate embryo the normal products of the ectoderm at the neurula stage are the cells of the non-neural ectoderm and the neural plate; at the tailbud stage they are the cells of the epidermis, the sensory placodes, the nervous tissue, the neural retina, the pigmented retina, etc., etc. The cataclysmic cell death that plays an important role in the modelling of many organs, and is bound up with the production of lytic enzymes, must also be considered as an achievement performed at a precise developmental stage. For instance, the area located in the posterior margin of the mesoderm of the chick wing bud that is called the posterior necrotic zone (PNZ), which can first be demonstrated at stage 17, undergoes necrosis at stage 24 [*Saunders and Fallon,* 1967].

Up to the present day the object of experimental embryology remains essentially to investigate to what extent the various parts of an embryo mutually contribute to the realisation of the normal achievements of each of them. To this end two main methods can be adopted, each of which has required the elaboration of various different techniques: (1) creating systematic deficiencies in the embryo, or isolating parts of it, allowing survival by culturing them in vitro or by in vivo culture in a cavity of another embryo (e.g. extra-embryonic coelom, anterior chamber of the eye); (2) modifying the environment of a rudiment either by grafting it to another region of the embryo or by associating it in vitro with other rudiments.

Of these methods the one that has proved to be most fruitful and is used most frequently is the in vitro culture of parts of embryos. Such explants generally continue developing for some time, sometimes even if they consist of only one cell type and the culture medium is a simple saline solution (in which case food reserves present in the cells must sustain their metabolism). They often assume the texture of a tissue or even the configuration of an organ characteristic of a later stage of development, something that is called self-differentiation. We prefer using the term 'autonomous progression of differentiation' [*Chandebois*, 1976 b], which does more justice to the dynamic nature of the process. Because most embryologists have been focussing their attention on the interdependence of the various parts of the embryo in morphogenesis they have given little attention to the theoretical importance of autonomous progression. However, it is fundamental to our understanding of the dynamics of differentiation and of the automatism of the reproduction of structural patterns.

The Autonomous Progression of Differentiation

> In a homogeneous cell population autonomous progression is an automatic process based on a continuous exchange of information (homotypic interactions) which is effected through cell contact relations. The programme for this autonomous progression is not communicated to the individual cells of the population at the time of determination. Rather, the programme assigns to the cells a certain elementary social behaviour, which is then the basis for the structuring of the population.

In typical cases a part of an embryo of homogeneous appearance, when isolated in a saline medium performs its normal achievement. When autonomous progression stops, its histological appearance is that of the tissue it would have produced, roughly *in the same period of time*, had it remained in the embryo. Moreover, parts of the same origin and initial size attain approximately the same ultimate size. For instance, the dorsal blastoporal lip of the amphibian gastrula, which represents the prospective notochord, develops into notochord when isolated. In the chick wing bud cultured in vitro the posterior necrotic zone undergoes necrosis at the same time as that of the contralateral wing bud left in place in the embryo [*Fallon and Saunders*, 1968]. More often, however, the result is not typical in that the cultured part performs the achievement of a different prospective area. For instance, an explant originating from the prospective somite area of the blastula does not necessarily form striated muscle but more often renal tissue or blood, which are normally produced by more ventral parts of the mesoderm [*Holtfreter*, 1938].

At the time the cells are engaged in their autonomous progression they evidently follow a prescribed 'trajectory'. They are propelled through a succession of transformations which can only proceed and stop on a rigorously prescribed time schedule when the cells have acquired a particular individuality. Obviously the nature of the progression is fixed at the start by the cytoplasmic information contained in the cells. As soon as the cells possess this information we say that the population is 'determined'.

The autonomous progression is not bound up with the individual activities of the cells alone: *it is a performance of the population as a whole.* If one disaggregates the cells of an explant capable of autonomous progression and cultures the cells in a dispersed state, the progression stops and is not resumed as long as the cells remain isolated [*Wilde*, 1961a]. When the cells reacquire the capacity of mutual adhesion the explant resumes its development more or less in the same way as an intact explant and the same specific substances become detectable. Thus, the cells of disaggregated mesoderm of the chick limb bud dedifferentiate and only synthesise chondroitin sulphate again after they have reaggregated [*Ede and Flint*, 1972]. Therefore autonomous progression is only possible thanks to the continuous exchange of information between the cells, which requires cell-to-cell contact through an intercellular matrix [*Wilde*, 1961a; *Elsdale and Jones*, 1963]. This exchange of information between like cells is known as 'homotypic interaction'. Experiments performed on the chick neural retina [*Moscona*, 1975] particularly clearly demonstrate the role of the cell surface in such interactions. During autonomous progression of retinal explants glutamine synthesis can be induced in the cells by hydrocortisone. Disaggregated cells lose this competence to respond to the hormone but reacquire it upon reaggregation. However, this is not the case if the cells are first treated with the lectin concanavalin A, which alters the properties of the cell surface.

Also the frequency of cell divisions during autonomous progression is a group effect. In small explants of limb mesoderm mitotic activity is higher than in large ones [*Flickinger,* 1976b]. Cells in general become mitotically more active upon dispersion.

Reaggregating dissociated cells spatially rearrange themselves in such a way that the tissue formed has a similar texture as the same tissue in the intact organism. This is particularly striking when the cells acquire different individualities because, for some reason or other, they communicate in different ways. Again the best example is furnished by chick limb mesoderm. In the intact embryo cell condensations appear in the centre of the young limb bud, which increase in size by the addition of more cells to their periphery. They develop into precartilage condensations surrounded by myoblasts [*Ede and Flint,* 1972]. The formation of the condensations is an expression of autonomous progression, for it occurs in the same manner in cultured mesoderm provided the amount of mesoderm is sufficient, even after disaggregation and subsequent reaggregation. The number of condensations is proportional to the amount of mesoderm; if only one is formed it lies in the centre of the mass. The precartilage condensations formed in vivo and in vitro have exactly the same histological appearance. Towards the periphery of the condensation cell adhesion apparently decreases: the cells are flatter and have a large radius of curvature [*Ede and Flint,* 1972]. The process of condensation explains why the mesoderm cells follow divergent pathways during autonomous progression. In fact, the cells can be pushed into different pathways by different culture conditions [*Flickinger,* 1976a]. If they are cultured on lens paper, on which they spread while increasing their mitotic activity, they form muscle cells. In contrast, if cultured in microdrops, which increases cell density and adhesion and reduces mitotic activity, they produce cartilage. In the precartilage condensations themselves the individualities of the cells change gradually: those in the centre stain more strongly for mucopolysaccharides than those at the periphery [*Ede and Flint,* 1972]. In the next chapter we will propose a model to explain this.

Other types of homotypic interaction may be involved in cell diversification in a cell population engaged in autonomous progression. In the amphibians ciliated cells appear in the superficial layer of the prospective epidermis at the neurula stage. They are interspersed among non-ciliated cells and one never finds two of them which are adjacent. Prospective ectoderm isolated from the 4-cell stage to the early gastrula and cultured in vitro develops in exactly the same way [*Grunz* et al., 1975; *Grunz*, 1977]. The cell interaction involved in this remarkable cell spacing phenomenon could be analogous to that suggested by *Wigglesworth* [1940] for the insect integument. The insect cuticle carries bristles (each produced by one cell) which retain the same spacing during growth of the integument. It seems as if a cell bearing a bristle inhibits the formation of other bristles within a certain number of cell diameters, possibly by absorbing some substance required for bristle formation. As the number of cells increases cells equidistant from neighbouring bristle-bearing cells will at a certain moment escape from their inhibitory range and form bristles in their turn.

Summarising, *autonomous progression seems to be an elementary automatic process in development; the programme for it is established at the time of determination.* This programme does not fix the destiny of each individual cell but lays down the rules for the exchange of information among the cells. In other words, *the event of determination imposes on the cell population as a whole a certain 'elementary social behaviour'* [*Ede*, 1972] *which automatically propels the cells towards new activities, determines their mitotic rhythm, makes them arrange themselves spatially in novel manners, and sometimes causes them to diversify their individualities.*

Embryonic cell populations isolated in vitro fail to produce adult tissues and only reproduce certain features of the structural patterns that are produced in vivo. This shows that a given autonomous progression covers only a relatively short period of the entire genesis of an organ and does not produce all the various cell individualities required for its formation. Information coming from other cell populations is needed to restart organogenesis and to enable certain cells to engage in differentiation pathways they cannot enter by themselves. This is what we call 'heterotypic action', in contrast to homotypic interaction. Heterotypic actions necessarily become more and more complicated as the properties of cell populations are transformed and diversified, and particularly when the tissues start to perform their respective functions in the organism.

Heterotypic Actions during Pre-Functional Organogenesis

The best known heterotypic actions are subsumed under the term 'induction'. They lead to quasi-immediate determination, in which one cell population, called the 'induced' population, irreversibly engages in a new differentiation pathway under the influence of a neighbouring population, the 'inducer', which itself usually is not transformed.

In the various different animal groups embryonic inductions take a large part in the organisation of the germ layers after they have been put in place during gastrulation [*Waddington,* 1956]. They are best understood in the vertebrates. The mesodermal archenteron roof (the prospective area of the notochord) induces the neural plate in the overlying ectoderm while it takes up its place during gastrulation. When contact between the two tissues is prevented no neuralisation of the ectoderm takes place. Various structures arising from the neural plate become inducers in their turn. For instance, various regions of the embryonic brain are required for the differentiation of the various cephalic placodes which later form the sense organs: the telencephalon induces the olfactory placode, the optic vesicle the lens placode and the rhombencephalon the ear placode. The segregation of the somites and the intermediate mesoderm from the lateral mesoderm depends on the influence of the median archenteron roof. When the dorsal blastoporal lip of a gastrula, which represents the chordomesoderm, is cut out and inserted into the blastocoele of another early gastrula, it first becomes pressed against the ventral ectoderm by the expanding archenteron and then gets incorporated into the ventral mesoderm. Here it induces somites and intermediary mesoderm, while it neuralises the overlying ventral ectoderm. Thus, an incomplete supernumerary embryo (usually the most anterior part) arises almost entirely from the tissues of the host embryo, only a portion of its notochord originating from the graft.

Later in development the somites again undergo induction by the notochord, as a result of which their sclerotomes produce cartilage. The intermediate mesoderm gives rise to the kidneys. Its cells disperse into mesenchyme and in the nephrogenic area a certain proportion of the cells condense into aggregates which form kidney tubules. In the anterior region this is an autonomous process that spontaneously produces the pronephros. From this the pronephric duct or primary ureter grows backwards and induces mesonephric tubules from the more posterior mesenchyme. In the amphibians the mesonephros is the definitive kidney, but in the birds and mammals a secondary ureter develops from the opening of the primary ureter and in the same manner induces the definitive kidney or metanephros from nephrogenic mesenchyme situated behind the mesonephros.

The Role of Induction in the Emergence of Tissues

> The inducer re-initiates or deflects the course of autonomous progression but it does not establish its programme. It selects one programme from among others which are already being executed. The execution of the programme – which under certain experimental conditions may continue or be accelerated in the absence of the inducer – consists in the stimulation of certain luxury metabolic strategies. The cells become competent to respond to the action of the inducer as soon as the inductive stimulus suffices to raise those strategies above a threshold beyond which they manifest themselves in visible structure.

People have always considered as evident the idea that induction implies the transfer of more or less specific substances from the inducer to the responding population. The identification of the putative inductive substances never appeared to pose serious problems. The methods of choice for this were (1) testing the effects on competent tissue of molecules extracted from the inducer, (2) tracing molecules transferred from the inducer to the induced population by radioactive labelling, or (3) sieving out such molecules by placing a millipore membrane filter between the two tissues. However, notwithstanding extensive use of these methods they have not led to conclusive results. Nevertheless, even today one often thinks that the role of the inducer is to activate in the cells of the induced population the genes that code for the proteins specific for the tissue to be formed. In present-day parlance one says that the cells are 'reprogrammed', that is, they are forced to use another part of their 'genetic programme'. In this view the global response of the induced tissue represents not more than the sum of the individual responses of the cells to the inducing substance. We will show here that analysis of a great many results that have accumulated in this area leads to the abandonment of this idea and to a quite different concept.

For many years embryologists have pointed out that one must first of all consider the properties of the induced cell population. The first reason for this is that the result of an induction depends more on the individualities of the responding cells than on the nature of the inducer. That the response rather than the inducer is specific becomes particularly clear in cases where several cell populations are transformed simultaneously by the same inducer. A case in point is the experiment in which a dorsal blastoporal lip is grafted into the prospective ventral mesoderm [*Spemann,* 1918]. The emergence of the supernumerary embryo is the result of three different inductions: (1) induction of a secondary neural plate in the ventral ectoderm; (2) induction of secondary somites and intermediate mesoderm in the lateral plate mesoderm; (3) induction of a secondary digestive tube in the yolk endoderm (induction of endodermal epithelium).

A second important reason is that the effect of an inducer can be mimicked by various other tissues called 'heterogenous inducers'. The most striking case is that of neuralisation, which can be obtained with almost any kind of inducer, be it embryonic or adult, living or dead, from vertebrates, invertebrates or even plants [review in *Dalcq,* 1960]. Moreover, all these heterogenous inducers can induce different tissues according to the individualities of the inducible cells. For instance, with spinal cord or salivary gland one can neuralise ectoderm or induce kidney tubules in nephrogenic mesenchyme.

Something that particularly underlines the importance of the individualities of the responding cells is the fact that they are only susceptible to the influence of the inducer during a very short period of development. This has been well demonstrated for the case of neuralisation of the ectoderm. If contact with the inducer is advanced by implanting a piece of dorsal blastoporal lip into the blastocoele of an early bastula the secondary neural plate appears only at the time the primary one emerges [*Mangold,* 1929]. If, on the other hand, contact is deferred by combining chordomesoderm with early gastrula ectoderm that has been withdrawn from mesodermal influences by keeping it as an explant in saline, neuralisation is only possible when the ectoderm was cultured for less than 48 h [*Holtfreter,* 1938]. Such experiments have led to the introduction of the *concept of 'competence'* [Waddington, 1932], *which expresses the fact that the capacity of an inducible cell population to respond to a stimulus in a quasi-immediate and specific manner is restricted in time.* Thus, the ectoderm of the early gastrula, during the brief period it responds to contact with chordomesoderm by neuralisation and to contact with lateral mesoderm by epidermal differentiation, possesses both neural and epidermal competence. Later, at the tailbud stage, the same ectodermal cells go through a period of lens competence. During these two consecutive periods the ectoderm does not respond in the same manner to the same heterogenous inducer. In the presence of a protein extract from bone marrow gastrula ectoderm forms brain structures flanked by optic vesicles, while tailbud-stage epidermis forms a lens [*Sasaki* et al., 1957]. The epidermis loses its lens competence when the normal lens placode has appeared.

In most inductive situations only part of the cells that are induced are in contact with the inducer. It is known that neural induction requires an intimate contact between the ectoderm and the archenteron roof. In vivo this contact is only established in the medio-dorsal region of the embryo. We must assume that neuralisation first occurs in that region and then extends laterally [*Nieuwkoop,* 1967; *Leussink,* 1970]. In vitro a piece of ectoderm brought into contact with a small piece of chordomesoderm is neuralised over a much larger area than the area of contact with the inducer [*Sala,* 1955]. Similarly, in nephrogenic mesenchyme brought in transfilter culture with an inducer a cell mass more than ten cell diameters thick participates in tubule formation, although only the cells actually touching the filter are in contact with the pseudopods sent out by the inducer [*Saxén and Saksela,* 1971]. Even today the propagation of induction is often attributed to the diffusion of one or more substances. However, such diffusion has never been experimentally demonstrated and some authors have argued that if it happens this probably means the tissue is dying [*Deuchar,* 1970a]. The hypothesis is the more doubtful since inducers themselves, at least in many cases, do not act in this manner, as we shall see later.

It is in better agreement with our present knowledge to ascribe the propagation of induction to a relay mechanism: within an inducible tissue each induced cell in turn induces the cells with which it is in direct contact. This has been unambiguously shown to be true in the case of the neural plate [*Mangold,* 1933; *Rasilò and Leikola,* 1976] but not in that of nephrogenic mesenchyme [*Saxén and Saksela,* 1971].

The cells that come directly or indirectly under the influence of the inducer do not respond individually: it is their elementary social behaviour that is modified. In other words, *the induced population as a whole is determined to follow an autonomous progression which it would be unable to engage in on its own.* A very good example is that of the kidney [*Saxén* et al., 1968]. When nephrogenic mesenchyme is associated in vitro with a tissue of epithelial origin (e.g. spinal cord) it forms aggregates which transform themselves into tubules, all this on a similar time scale as in the intact animal. The metabolic activities of the cells develop in the same manner: in particular, proteins specific for the adult kidney appear after 12 days. All these changes can proceed in the absence of the inducer as soon as the responding tissue has been determined, that is, after a period of contact between the two tissues of about 8 h. But if once determined mesenchyme is disaggregated it stops developing. It only resumes its development (after reaggregation) if it is once more subjected to the influence of the inducer. Therefore, the influence of the inducer does not definitively transform each individual cell. Rather, their social behaviour as a cell group is irreversibly modified as long as they remain together – notably the adhesiveness of the cells as it manifests itself in the formation of tubules.

Neuralisation also is clearly a response of a whole population of cells. In vitro it does not occur if the piece of responding ectoderm does not consist of a sufficient number of cells [*Deuchar,* 1970b, c; *Johnen and Albers,* 1978]. Determination, which in this case only requires a few minutes of contact, is the starting point for an autonomous progression involving homotypic interactions. For instance, cells of the neural crest which are disaggregated, washed and cultured in vitro fail to differentiate into pigment cells if their number is insufficient (less than 15), unless intercellular material is produced again by the cells [*Wilde,* 1961b]. The event of determination also has distinct effects on the other two aspects of the social behaviour of induced cells that are of great importance for the understanding of morphogenesis. First, it immediately stimulates mitotic activity, so that the cells of the neural plate come to lie in several layers. Secondly, it confers on the cells the capacity to arrange themselves spontaneously around a central cavity and to place themselves inside other tissues. This becomes particularly evident when the ectoderm of the advanced gastrula is disaggregated (which in this case does not efface their neural determination). During reaggregation the neuralised cells end up in the centre of the aggregate where they form a hollow sphere completely surrounded by epidermal cells. On sections this neural vesicle looks exactly like the early embryonic nervous system [*Townes and Holtfreter,* 1955].

In certain cases the inducer deflects the course of the autonomous progression in which the cells are engaged, and in which they would have persisted in its absence. The cells are thus shunted into another pathway of differentiation. The isolated animal half of the blastula entirely develops into ectoderm, although it comprises part of the prospective mesoderm. It only forms mesoderm if associated with yolk endoderm [*Nieuwkoop,* 1969a]. Apparently the future mesodermal blastomeres are at first engaged in an autonomous progression leading to the production of ectoderm – and to the ultimate acquisition of neural competence. However, in this progression the cells first acquire mesodermal competence, and this manifests itself under the influence of the yolk endoderm, by which their autonomous progression is deflected.

Another example is the somite, which comprises three prospective areas: that of the myotome, which will produce striated muscle, and those of the dermatome and sclerotome, the cells of which will disperse and form dermis and cartilage respectively. An early somite in culture entirely develops into muscle. To make it form cartilage it must be associated with a piece of notochord for at least 12 h. Cartilage later appears in the region of contact, even if this is the prospective myotome or dermatome [*Grobstein and Holtzer,* 1955]. Yet another example is that of the optic vesicle. Normally this differentiates into neural and pigmented retina. When cultured as an isolated explant in saline it develops entirely into neural retina, but when it is surrounded on all sides by mesenchyme it produces pigmented retina only [*Lopashov and Stroeva,* 1961].

In other cases induction re-initiates autonomous progression after it has 'run out'. Here the emergence of various tissue types from the same inducible cell population requires the action of several inducers. This is well illustrated by the differentiation of the ectoderm in amphibians. If one explants prospective ectoderm taken from the late blastula one usually obtains neither neural structures nor epidermis or epidermal derivatives. After some time the cells more or less dissociate and then die. The ectoderm behaves in the same way in the case of exogastrulation, when unsuitable environmental conditions (such as a non-balanced saline solution [*Holtfreter,* 1933]) prevent proper gastrulation and thereby the interiorisation of the mesoderm and endoderm. Ectoderm isolated from the gastrula stage does not develop beyond that stage. It is only in contact with the archenteron roof that it becomes neuralised, and only in contact with lateral mesoderm that it is determined as prospective epidermis. If one for instance transects the blastula at the level of the equator and re-associates the two halves after 180° rotational transposition, normal development ensues even though now the archenteron roof ends up under the ventral ectoderm. Consequently the latter now forms the neural plate, and the neurectoderm common epidermis [*Spemann,* 1918]. All these facts can be summarised by saying that the inducer effects a choice from among various histogenetic potencies.

A tissue that usually manifests a given competence by engaging in an autonomous progression under the influence of an inducer can sometimes do this without the inducer. For instance, ectoderm taken from an axolotl gastrula before having been in contact with mesoderm is partially neuralised in saline medium [*Barth,* 1941], whereas similar ectoderm from a frog gastrula in similar conditions can form epidermal glands [*Nieuwkoop,* 1963]. In the axolotl the mesonephros can develop in the absence of the pronephric duct [*Yamada,* 1940; *Burns,* 1955]. Erythrocytes normally originate from the lateral plate mesoderm. In the chick embryo (as in the amphibian embryo) the determination of the mesoderm requires induction from the endoderm during the cleavage stages [*Eyal-Giladi and Wolk,* 1970]. However, non-incubated blastoderms have been cultured in serum-free medium in such a way that this induction does not take place, and erythrocytes were formed nonetheless [*Zagris,* 1980]. We may also mention that posterior somites of the chick embryo can form cartilage in the absence of notochord [*Lash,* 1967]. Because it occurs more readily in cultured tissues, it seems that such spontaneous manifestation of a competence is simply related to a change in the contact relations between the cells. Undifferentiated optic vesicles of the chick embryo placed in suspension culture develop entirely into neural retina, which is characterised by intimate cell contacts leading to a prismatic epithelium. In contrast, if the cells are allowed to spread on a substratum, cell contact is weakened and the cells form pigment [*Dorris,* 1938]. Similarly, flank mesenchyme can behave like limb bud mesenchyme, forming precartilage condensations in vitro, but only after it has been previously dissociated [*Crosby,* 1967]. The dispensability of the inducer in certain species and certain experimental conditions shows that *the cells of a competent population already prior to determination possess the metabolic machinery needed to undertake their autonomous progression.* Because this same tissue can sometimes also perform other achievements in the absence of an inducer – or under the influence of other inducers – other metabolic machineries must be present that can initiate other autonomous progressions [*Ellison and Lash,* 1971]. In other words, *the inducer does not furnish information needed for the establishment of the programme of the autonomous progression. It selects one out of two (or more) programmes already available at the time of induction.* This idea is corroborated in a striking manner by observations made by *Wilde* [1961 a, b], which apparently ran so much counter to current concepts that they have not received the attention they deserve. Embryonic cells which are disaggregated and then washed, slowly reform the intercellular material that has been removed, and then redifferentiate. It sometimes happens that one cell differentiates in two different directions at the same time: part of the cytoplasm may contain melanin and another zone muscle fibres.

If we assume with the majority of present-day biologists that induction leads to the activation of the genes specific for the tissue to be induced, this concept, *as well as the experimental facts that support it,* would be devoid of any meaning. However, if one accords general purport to the facts revealed by the study of the transdifferentiation of the optic vesicle into lens (p. 17) one can a priori advance a different hypothesis. Histogenetic potency appears to be bound up with the maintenance in latent form of a certain luxury metabolic strategy; when it becomes manifest in a structural alteration of the cells this implies a stimulation of the strategy in question. We can then predict that the specific substances found in the induced tissue must have been synthesised in small amounts already prior to induction, not only in the prospective area of the induced tissue but also in all those anlagen that possess the potency to give rise to the tissue concerned under various experimental conditions. And in fact in the literature of the last 30 years we find numerous data which support this idea [*Chandebois,* 1981].

For various tissues it has been ascertained that *the specific proteins do not appear de novo at the time of morphological differentiation: they can already be detected, though in much lower concentrations, in the prospective area of the tissue.* For instance, two rudiments of the chick embryo which will later form both muscle and cartilage already produce myosin and chondroitin sulphate at the undifferentiated stage: the epithelial somite [*Lash,* 1968; *Orkin* et al., 1973] and the early limb bud [*Searls,* 1965; *Medoff,* 1967; *Medoff and Zwilling,* 1972]. In the latter case, at the time of muscle differentiation the amount of myosin increases (50-fold in 24 h [*Medoff and Zwilling,* 1972]), while the amount of the corresponding mRNA that is being translated on polyribosomes rises [*Dym* et al., 1979].

Moreover, substances specific for certain tissues can often be detected already at very early developmental stages, *even outside the prospective areas of the tissues in question.* Globin [*Wilt,* 1962], α-actin [*Ballantine* et al., 1979; *Sturgess* et al., 1980], myosin [*Orkin* et al., 1973; *Ranzi,* 1980] and chondroitin sulphate [*Franco-Browder* et al., 1963] have been detected at the very earliest stages of amphibian and avian development; the same holds for the moulting hormone of insects [*Hsiao and Hsiao,* 1979]. Lens-specific proteins [*Flickinger and Stone,* 1960; *Perlmann and De Vincentiis,* 1961] and contractile heart proteins [*Ebert,* 1953] have been found in the posterior portion of embryos, although the corresponding organs develop only much later in the anterior portion. Apparently, therefore, the various parts of the embryo simultaneously maintain several luxury metabolic strategies. In a given area only one of them will ultimately be promoted to the dominant status. The others persist in a latent state, at least for some time. Crystallins are found both in the chick optic vesicle (between the 3rd and 8th day of incubation [*Clayton* et al., 1979]) and in the chick adenohypophysis rudiment [*Clayton,* 1979]. In the rat insulin has been detected in almost all tissues of the fetus and newborn [*Clark and Rutter,* 1967]. Shorter or longer after determination latent metabolic strategies can be revived under exceptional experimental conditions. For instance, the ventral epidermis, which seems to lose its neural potency after the formation of the neural plate, can still transdifferentiate to nervous tissue after the tailbud stage [*Takaya,* 1977]. In parallel with the presence of crystallins and the corresponding mRNAs, the optic vesicle of the chick embryo for some days retains the capacity to transdifferentiate into lens [*Clayton* et al., 1979] (see also p. 17). Similarly, in parallel with the presence of glycosaminoglycans in the mesonephric rudiment [*Abrahamson* et al., 1975] and the lateral plate mesoderm [*Marzullo and Lash,* 1967; *Stephens* et al., 1980], cartilage can be produced by the embryonic kidney in vitro [*Lash,* 1963] and by the flank mesoderm when it behaves like limb mesoderm after dissociation [*Crosby,* 1967].

Summarising, it is now clear that we can attribute a general significance to the concept of induction once proposed by *Holtzer* [1952] for the specific case of cartilage: *the role of the inducer is not to establish specific new activities but to stimulate activities that had already been established previously, while restricting their localisation in the embryo.* It does so by modifying the properties of cell surfaces, for which in vitro sometimes a simple change in the ionic environment of the cells [*Barth and Barth,* 1974] or simple mechanical effects are sufficient. In vivo such modification necessarily requires contact between the cells of the inducer and the induced cell population. This inductive mechanism is particularly well exemplified by the induction of the kidney in the mouse. In vitro, kidney tubules can be obtained by placing nephrogenic mesenchyme on one side of a millipore membrane filter and some epithelial tissue (spinal cord or salivary gland) on the other side. Electron microscopic examination shows that the filter pores contain cytoplasmic processes of the inducing cells rather than extracellular material [*Wartiovaara* et al., 1974]. That the cell processes are responsible for the induction is shown by the fact that no aggregation of the mesenchyme occurs when the diameter of the filter pores is smaller than that of the processes sent out by the inducing cells [*Saxén* et al., 1976]. It seems that a reduction in the synthesis of glycosaminoglycans at the level of the interface between inducer and target cells is involved in the induction [*Ekblom* et al., 1979].

It is a well-known fact that substances extracted from certain tissues can effect induction. However, the experimental conditions under which this has been observed (particularly culture in vitro) have always been such that the internal cohesion of the induced tissue was more or less altered. Also, the fact that the induced rudiment is always clearly delimited makes it difficult to maintain the idea that inducers could be diffusible substances [*Toivonen* et al., 1976]. Nevertheless, certain data obtained on the differentiation of cardiac muscle cells in the chick are very interesting. It can be evoked in the chorio-allantoic membrane by injection of extracts of differentiated cardiac muscle mixed with Rous sarcoma virus [*Ebert,* 1959], and also in the prospective heart area by short dRNAs extracted from the same material [*Niu and Deshpande,* 1973]. These molecules are not mRNA, for they are not translatable [*Deshpande and Siddiqui,* 1977]. One could suggest that this specific RNA is a 'primer'. When forced by the experimental conditions to enter cells which already synthesise a small amount of the same molecule, it would stimulate the transcription of genes coding for heart-specific proteins and thus raise it above the threshold beyond which morphological differentiation becomes manifest.

In understanding induction the nature of the signal emitted by the inducer is not the crucial question: we must go back into the past of the inducible cell population. In certain cases the acquisition of competence is clearly the outcome of an autonomous progression. The clearest case is that of neural competence, which spontaneously appears in prospective ectodermal blastomeres isolated from the blastula. In general, such an autonomous progression can be modulated, for the manifestation of competence in vitro can be advanced to a greater or lesser extent depending on the technique used. For instance, in chick somites cultured in association with notochord, cartilage appears earlier than in normal development [*Lash,* 1967]. Because we know that autonomous progression implies continuous homotypic interactions mediated by cell-to-cell contacts, it seems probable that the temporal variation in the appearance of competence (and in its spontaneous manifestation) is simply bound up with changes in cell adhesion resulting from the nature of the substratum. For instance, chick somites explanted at stage 9 are unable to form cartilage under the usual culture conditions but can produce cartilage if cultured (in the presence of notochord) on a film of mineral oil [*Ellison* et al., 1969].

In other cases heterotypic actions are required for competence to appear, for instance in the case of the sensory placodes [*Jacobson,* 1963, 1966]. In the newt *Taricha* the placodal ectoderm does not form placodes in vitro. If one associates it systematically with other tissues which in the embryo adjoin it permanently (neural plate) or only temporarily (endoderm, cardiac mesoderm) one always obtains placodes. Therefore, morphological differentiation of a placode is not the result of a single factor but of a cumulative stimulation of the cells by various tissues which all 'do the same thing', including the last in the series, which is usually considered to be the natural inducer. Here again holds that if the cells are manipulated in such a way (particularly by culturing) as to react more strongly to any one of such a series of factors with similar

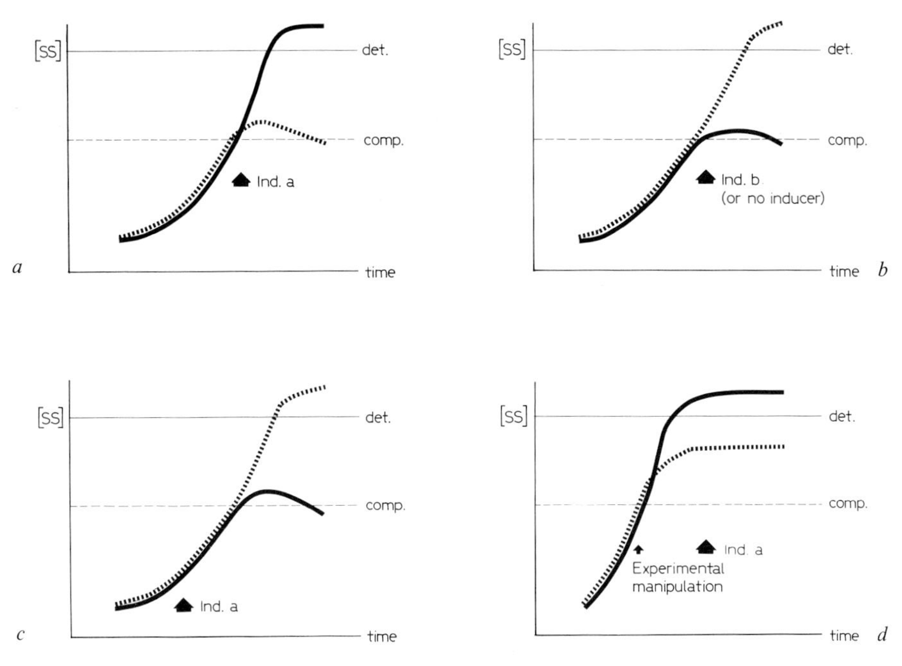

Fig. 3. Theoretical scheme of the modification of luxury metabolic strategies during tissue differentiation. [SS] = Concentration of specific substances. (——) = Strategy destined to become dominant under the influence of a given inducer (Ind. a). (– – –) = Other strategy destined to become dominant under the influence of a different inducer (Ind. b) or in the absence of an inducer. comp. = Threshold beyond which the cells become competent to respond to a natural inducer. det. = Threshold beyond which the cells differentiate (i.e. manifest visible structure). *a, b* Stimulation of different strategies by different inducers. *c* If the inducer acts on cells that are not yet competent it has no effect. *d* Early modification of homotypic interactions by experimental manipulation has the same effect as Ind. a would have; the inducer is no longer necessary.

effects [*Karkinen-Jääskeläinen,* 1978] morphological differentiation may manifest itself without the influence of the others; particularly the last one in the series, which in the usual experimental circumstances appears to be the actual inducer, can then be dispensed with.

In conclusion, *the programme of the autonomous progression that is singled out by an inducer was already being executed before that time* (fig. 3 a, b). *All the inducer does is to complete that which was begun by homotypic interactions or heterotypic actions, or both.* The inducer raises certain luxury metabolic strategies (one only in the case of an induction leading to terminal differentiation), which were already *stimulated* previously, to a level that exceeds a threshold beyond which they manifest themselves in visible structure. The inducer does not act if the autonomous progression has not advanced far enough (fig. 3 c); it is useless if the progression is too far advanced (fig. 3 d). A simile may further clarify this concept: induction may be compared to pouring a certain amount of water into a vessel that already contains water. This will be insufficient to fill up the vessel if the amount of water already present is too small; if, on the other hand, the vessel has already overflowed the addition of more water has no effect whatever. The major importance of this novel concept is that it enables us to explain by one and the same mechanism the three phenomena that are involved in the genesis of a given tissue type: induction, autonomous progression and transdifferentiation. In all three cases it is the modification of cell contact relations, related to the social behaviour of the cells or to experimental manipulation, which results in the elevation of one or more latent luxury metabolic strategies to the rank of dominant strategy.

The Role of Induction in the Diversification of Cell Individualities

The spatial demarcation of the induced rudiment – and sometimes its structuring and that of its surroundings – is not just an imprint left behind by contact with the inducer. This is because of (1) the social behaviour of the cells (i.e. propagation of induction, modification of cell contact relations), and (2) the fact that the stimulation of a luxury metabolic strategy destined to become dominant is a progressive process.

The result of induction is not only the emergence of a new tissue. It is also the appearance of the rudiment of an organ or organ system with accurately fixed boundaries. In the majority of cases it is quite clear that the induced area is not an imprint left by the inducer, as would be the case if all cells responded individually to the induction. There is no strict relationship between the size or shape, or both, of the induced structure and the inducer.

The often distinct boundary between the induced area and the area escaping induction can be explained by the progressive loss of competence from the inducible cell population (fig. 4 a). Thus, after a certain time during which it spreads across a definite number of cell diameters, the inductive influence would no longer evoke a response. This mechanism obviously can-

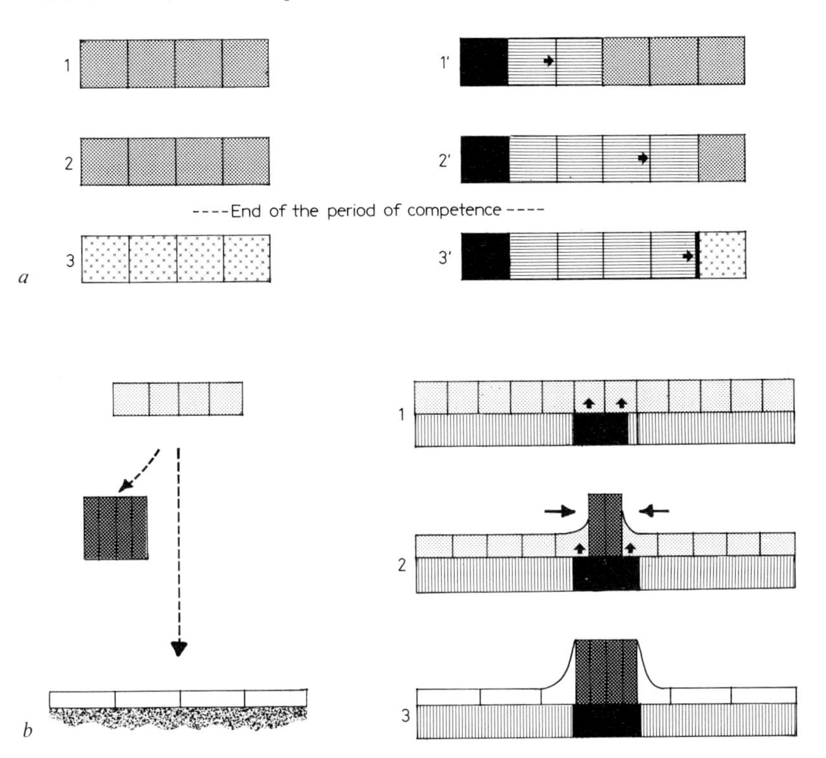

Fig. 4. Theoretical models to explain the exact positioning of the boundaries of induced rudiments. *a* The propagation of induction stops because competence is lost in the induced tissue. Left: Autonomous progression in the absence of the inducer. Right: Autonomous progression in the presence of the inducer. Crosses: non-induced tissue; dots: competent tissue; hatched: induced tissue. *b* Delimitation by the effect of traction due to a change in adhesiveness of the induced cells. Left: In vitro the tissues can differentiate in two different directions according to whether the cells do or do not spread on the substratum. Right: In vivo or in vitro a number of cells which adhere more strongly to each other as a result of induction exert sufficient traction on the adjoining cells to force them into another differentiation pathway.

not be involved in cases where the size of the induced structure is smaller than that of the contact area with the inducer, for instance when a lens is determined by endoderm and mesoderm in the absence of the optic vesicle [*Jacobson*, 1966] or by a larger optic vesicle from a different species [*Balinsky*, 1957]. The same holds for cases where determination occurs spontaneously in vitro and only involves part of the explanted tissue. If we remember the numerous studies showing that simple physical factors that bring about spreading of cells on a substratum can influence the progression of their differentiation, a different hypothesis can be proposed (fig. 4 b) (which does not necessarily exclude the first one). When induction brings about an increase in cellular adhesiveness (for instance in sensory placodes) this will result in a local contraction or condensation of the tissue, the adjoining cells being subjected to traction, the resultant depending on their adhesion to neighbouring tissues. When induction has extended over a certain number of cells (which is always the same in intact embryos of the same species but variable under experimental conditions) the traction becomes strong enough to force the adjoining cells to release their contacts and consequently to enter a different pathway of differentiation. This hypothesis finds its justification in a reconsideration of experiments carried out by *Nieuwkoop* [1963]. When two pieces of competent axolotl ectoderm are cultured as a sandwich (that is, joined by their inner surfaces and without inducer) 'spontaneous' neuralisation only occurs along the periphery, probably under the influence of the medium. This is succeeded by a contraction of the outer rim of the sandwich, compensated by stretching of the central part, which develops into ordinary epidermis. When competent ectoderm is cultured singly on a collodion film with the internal surface downward, a flux of cells occurs towards the periphery, where the tissue is attached to the collodion, and it is there that islands of neuralised ectoderm will be formed.

In an area where induction occurs one usually sees a sudden increase in complexity of organisation, which reflects a diversification of cell individualities both in the induced cell population and in the adjoining tissues. This is particularly puzzling in the early germ layers, where structures seem to emerge ex nihilo. It cannot always be attributed to some pre-existing, invisible organisation in the inducible tissue simply made manifest by morphological differentiation, nor to the organisation of the inducer imprinting itself on the induced tissue. A case in point is the artificial neuralisation of axolotl ectoderm in saline solution. Here well-identifiable prosencephalic structures are formed in the absence of an inducer [*Barth,* 1941]. However, we should not forget that determination confers a specialisation on the cells that amounts to a 'theme' on which many variations are possible (p. 21). Throughout the autonomous progression that succeeds determination the cells continue to communicate with each other and with those of neighbouring cell populations. *Cells which are identical at the start can only develop in exactly the same manner if they receive the same extracellular information at the same time.* Because this situation is never realised cell types diversify after induction, even if the cell population concerned is strictly homogeneous at the beginning.

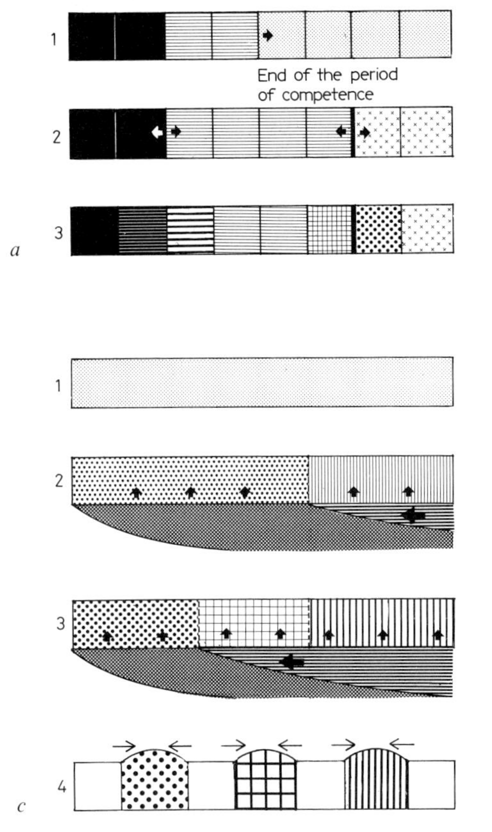

Fig. 5. Theoretical models to explain the diversification of cell individualities in an induced tissue. *a* Exchange of information between the induced (hatched) and the non-induced tissue (crosses) on the one hand, and between the inducer (black) and the induced tissue on the other. *b* Progressive induction. *c* Induction of three types of tissue by only two inducers; explanation in text.

The first cause of diversification is that the cells of the induced population do not all occupy the same position with respect to the inducer and the non-induced tissue (fig. 5 a). Cells finding themselves on either side of the boundary between the induced and the non-induced tissue will exchange information and consequently differentiate in different manners. This situation is exemplified by the neural plate and its adjoining tissues. The neural crest is formed by cells which were first neuralised and are then transformed under the influence of the adjacent non-neuralised ectoderm. The ectoderm in the vicinity of the neural plate is in its turn weakly neuralised and gives rise to the placodal ectoderm [*Nieuwkoop,* 1963]. The same holds for the cells on either side of the boundary between the inducer and the induced tissue where 'two-way induction' occurs. For example, after the yolk endoderm has induced the mesoderm from the prospective ectoderm of the blastula, it itself needs the influence of the mesoderm to form endodermal epithelium [*Nieuwkoop,* 1969a]. We may also mention here that the contact that is established between the optic vesicle and the epidermis leads not only to the appearance of the lens placode but also keeps a considerable portion of the optic vesicle from spreading on the surrounding mesenchyme, as a result of which that portion forms the neural retina [*Lopashov and Stroeva,* 1961].

The transmission of the inductive influence from one cell to the next requires some time, so that the cells are determined at different phases of their current autonomous progression (which ends with the loss of competence). We may therefore expect different cell individualities to result. We will return to this point later.

Another cause of diversification of cell individualities in the induced tissue resides in the fact that overt differentiation is the end result of gradual and progressive stimulation of one or more luxury metabolic strategies. The outcome may vary if the heterotypic action involved in determination does not last equally long everywhere, or if several heterotypic actions all acting in the same sense vary spatially. The first case (fig. 5 b) is exemplified by experiments carried out on the mesoderm [*Yamada,* 1940]. We have already seen that grafting a piece of chordo-mesoderm to the ventral mesoderm leads to the formation of somites and intermediate mesoderm, giving rise respectively to striated muscle and kidney (p. 31). Prospective somites isolated at the middle neurula stage form kidney, while isolated prospective intermediate mesoderm forms blood, the normal achievement of lateral plate mesoderm that falls outside the range of influence of the notochord. If the same tissues are cultured in association with notochord, however, they both form somites. Obviously in the normal embryo the prospective somite tissue is first determined to form kidney; the continued presence of the notochord is required for it to form muscle. In this so-called 'progressive' type of induction the action of the inducer is prolonged and leads to a series of successive determinations. At the time competence disappears only the cells closest to the inducer will have gone through the whole range of determinations.

The second case (fig. 5 c) was suggested to us by the emergence of the three sensory placodes (olfactory, lens and auditory) from the same placodal ectoderm. They are already specifically determined at the open neural plate stage, i.e. prior to their final induction by the brain. Placodal ectoderm isolated from the early neurula can form placodes in the presence of endoderm or cardiac mesoderm, or both. The olfactory placode is determined exclusively by endoderm, the auditory placode by mesoderm, and the lens placode by either of the two [*Jacobson*, 1963]. Thus, the combined action of those two tissues brings about the differentiation of three types of placode. If the same experiments are done with isolated ventral ectoderm determination requires the influence of the telencephalon, the optic vesicle or the rhombencephalon, respectively (see also p. 57).

Heterotypic Actions during Functional Organogenesis

The establishment of the various tissue functions early in development does not mark the end of the differentiation process. The specific substances involved often still do not have their definitive composition. For instance, fetal haemoglobin is not identical to adult haemoglobin. Moreover, the diversification of cell types continues – including cell death, which still plays a large part in organ morphogenesis. It is reflected not only in the late determination of certain anlagen but also in the appearance of minute structural detail. Also in this second phase of development the progression of differentiation requires heterotypic actions, but these are more diverse in nature because certain tissues as they start functioning acquire new properties, those of the adult form. On the one hand, there is wearing-out of cells, which requires cell replacement from stem cells. On the other hand, the endocrine glands mature and start producing hormones, which the newly established vascular system carries through the whole organism.

Cell Renewal

> In cell populations undergoing renewal heterotypic actions are required for the maintenance of the stem cells and, in certain cases, for the emergence and maintenance of certain specific tissue activities. In cases where the tissue consists of cells at various stages of their autonomous progression, renewal contributes to the amplification of the diversity of cell individualities.

In tissues subject to renewal, stem cells (generative cells) are kept in reserve as a result of inhibition exerted by an adjoining tissue; this temporarily blocks the autonomous progression into which they had been shunted by a prior determinative event. Probably each time a stem cell divides one of the daughter cells continues to be inhibited and remains a stem cell, while the other daughter cell escapes from inhibition, completes its differentiation, and then wears out and dies. This is suggested by experiments on embryonic chick skin cultured in vitro [*MacLaughlin,* 1963]. If the epidermis is cultured entirely without its adjoining connective tissue all cells produce keratin and die within 10 days. Similar experiments with adult mouse skin have shown that the same mechanism is still at work in the adult.

If the cells of the generative layer still possess more than one potency, each time a cell engages in its definitive differentiation a specific extracellular factor may deflect its autonomous progression into a different pathway. It follows from this that the maintenance of certain tissues subject to renewal permanently requires information originating from another tissue. It follows also that certain tissues indefinitely remain competent to certain factors from other tissues. When subjected to one of these factors such a tissue changes its nature and function: it undergoes 'modulation'. Unlike transdifferentiation, where all individual cells of a tissue are transformed, modulation occurs thanks to the replacement of worn-out cells by cells of another type. For instance, when embryonic chick epidermis, which normally produces keratin, is isolated from the underlying mesenchyme and recombined with connective tissue of various origins or grafted to the conjunctiva of the eye, it can differentiate into mucous, ciliated or endothelium-type epithelium depending on the origin of the connective tissue. In adult mammals already keratinised epidermis can be transformed, often reversibly, into various kinds of glandular epithelium [references in *MacLaughlin,* 1963].

In a tissue subject to renewal the various cell types observed do not necessarily represent lineages that in the past have registered different sorts of extracellular information. They may at least in part represent successive phases in the transformation the cells undergo as they are displaced by younger cells towards the region where they will finally be destroyed. A typical example is that of the hepato-pancreas of the crayfish. Its glandular tubules grow from stem cells at the tip while cells die at the base. During their displacement they of necessity form absorbing cells, secretory cells and fibrillar cells [*Davis and Burnett,* 1964]. This mechanism of differentiation also seems to be utilised by the vertebrates. An example is the differentiation of the various regions of the urinary tubules of the kidney [*Burnett and Haynes,* cited in *Burnett,* 1967]. One can see how important cell renewal must be in the reproduction of structural patterns. However, the stem cells which characterise renewing tissues are not always easily detectable histologically, and in general experimental work on their properties is still very scarce.

The special case of renewing tissues has great theoretical importance because it shows that *not all heterotypic actions involved in the completion of the differentiation of functional tissues are comparable to embryonic inductions. Certain heterotypic actions temporarily block autonomous progression so that it will not be completed before its appointed time, while others are continuously required to have permanent effects.*

The Role of Morphogenetic Hormones

> Morphogenetic hormones play a role analogous to that of the heterotypic actions occurring between neighbouring tissues. Although they circulate through the whole organism they nevertheless contribute to the diversification of cell types by modifying the exchange of positional information among cell populations.

When the progression of differentiation and cell diversification comes to a stop (which normally happens at the end of organogenesis) this means that cell interactions have reached a state of equilibrium. In animals which possess a circulatory system this state would be reached before the end of development, were it not for the action of the morphogenetic hormones. As soon as the endocrine organs become functional these hormones are carried through the entire organism, bringing new information to the cells and transforming the individualities of many of them. A certain imbalance among the tissues is the result, and the progression of differentiation starts anew so that cell diversification proceeds. Thanks to the hormones development gets its 'second wind'. This is particularly evident in animals like the frog, where organogenesis, which had been more or less 'marking time' after hatching, is finished during a complete remodelling of the organisation of the tadpole. It is a single hormone, thyroxine, that is responsible for the whole metamorphic process [*Etkin,* 1935]. It advances it when injected into young tadpoles, but when during this period the thyroid gland is extirpated the resumption of organogenesis is blocked [references in *Dent,* 1968].

In many cases the influence of a morphogenetic hormone is equivalent to a determinative event. The extirpation of the thyroid of a frog does not lead to the reappearance of larval features. Moreover, the various tissues respond according to their current identity, and therefore in the most varied ways. Some tissues, which had remained undifferentiated after induction, now complete their differentiation. A case in point is the mesonephros, which takes over from the pronephros during metamorphosis. Other tissues, which were already functioning in the larva, change their specific synthetic activities; for instance, the purple pigment in the retina is modified through substitution of rhodopsin for porphyropsin [*Ohtsu* et al., 1964; *Weber*, 1967]. In certain places (e.g. tail, branchial apparatus and circulatory system of anurans) the changes involve the new synthesis of RNA and proteins (enzymes) that lead to cytolysis [*Tata*, 1966].

To be responsive to the action of a hormone a tissue must have achieved a certain degree of maturity. For example, thyroxine injection remains without effect in an early tadpole that still has external gills [*Etkin*, 1950]. This type of competence is typically of long duration. In the axolotl, where metamorphosis only occurs occasionally and as it were by accident (the adult usually more or less retaining the larval form) it can still be evoked by thyroxine injection, at least until sexual maturity is reached.

Not all hormones involved in the completion of differentiation have irreversible effects comparable to those of embryonic inducers, or if they have them they do not have them on all tissues. Just as is the case for the heterotypic actions between topographically adjoining tissues in this period, there are cases where a hormone first calls forth a new tissue activity and is then required for its maintenance, and other cases where it temporarily blocks the autonomous progression of differentiation until it is completed at the proper time. These two modes of action are exemplified by the sexual morphogenetic hormones. When the embryonic gonadal rudiments are removed only one type of genital tract develops, that of the so-called neutral sex, whatever the genetic sex of the embryo, whereas the genital tract of the so-called dominant sex remains rudimentary. Therefore, the hormone produced by the gonads of the dominant sex inhibits the differentiation of the genital tract of the neutral sex. On the other hand, the temporary disappearance of certain secondary sex characteristics in the period when the gonads are inactive clearly shows that their appearance and maintenance are under the continuous control of the sex hormones [references in *Gallien*, 1973; *Hogart*, 1978]. Another example is that of metamorphosis in higher insects, which is triggered by a drop in the level of juvenile hormone and thus in part represents the removal of an inhibition: juvenile hormone keeps the imaginal discs (the rudiments of the adult organs, which are determined in the embryo) in an undifferentiated state during larval life [references in *Doane*, 1973].

If one only considered the reactions of individual cells one might be led to think that a morphogenetic hormone, circulating throughout the organism, could not contribute to the diversification of cell individualities. In actual fact, the transformation of the individuality of any cell in response to a hormone necessarily has repercussions for the neighbouring cells because they no longer receive the same positional information. This is shown most clearly by certain tissues which are themselves not responsive to a particular hormone but nevertheless undergo a transformation when they are associated with another tissue in the presence of the hormone. An example is the epidermis of the chick embryo, which behaves in two different ways if cultured in a medium containing oestradiol: when cultured alone it keratinises, as if no hormone were present, but when cultured in association with oviduct mesenchyme it produces mucus [*Moscona,* 1961]. It follows from this that under the influence of a hormone a tissue may become an inducer and determine the emergence of a new organ from an adjoining tissue. A case in point is the visceral skeleton of the anuran tadpole: under the influence of thyroxine its elements give rise to the bones of the middle ear, and these in their turn induce the tympanic membrane in the overlying skin, and this happens even if a piece of flank skin is grafted over the ear region [*Helff,* 1925].

The transition from the phase of functional organogenesis to the adult state does not bring with it profound changes in the exchange of information. We can say of any cell that a smaller or larger proportion of the visible or invisible features of its individuality are not irreversibly determined even in the adult – for those cells of certain lower invertebrates which can re-acquire totipotency this even holds for all the features of their individuality. Consequently, *even when development as such has reached its end point the homotypic interactions and those kinds of heterotypic action that have reversible effects do not cease: they reach a stable equilibrium that maintains the state that has been acquired.*

The Summation of Extracellular Information and the Effect of Position

> Autonomous progression must be considered as the prime mover of development. During the differentiation of a cell lineage the succession of heterotypic actions – i.e. the registration of extracellular information, the sum of which determines the ultimate individuality of the cell – represents a sort of 'remote control' that may re-start, deflect or stop autonomous progression.

If we trace the genealogy of any cell of the organism back to the egg, we find that the progression of differentiation in its 'ascendant line' is continuous and – at least for the greater part – consists of quantitative transformations. At the beginning all the luxury metabolic strategies maintain themselves at very low levels of activity – which is reflected in histogenetic totipotency. From among these various strategies only one is gradually elevated to the status of dominant strategy. The others remain latent, sometimes after passing a period of stimulation. They may either stay latent indefinitely or may ultimately be abandoned altogether. All this leads to the progressive restriction of histogenetic potency (fig. 6 a). These changes in the relative activity of luxury metabolic strategies is well illustrated by the case of chondroitin sulphate in the frog [*Kosher and Searls*, 1973] and the chicken [*Abrahamson* et al., 1975]; at the beginning of development it is found in a large variety of tissues, particularly the epidermis, the endoderm and the neural tube, but it later disappears altogether from the latter while its concentration remains high in the notochord and particularly in the sclerotomes (the cartilage-forming portions of the somites).

New visible histotypic traits appear in the individualities of the cells of a differentiating lineage whenever certain specific substances exceed a certain concentration threshold. However, current techniques fail to detect these substances when they are present in insufficient amounts. This is why most embryologists have acquired and maintained a concept of differentiation that is diametrically opposed to the one we are proposing. They view the progression of differentiation as a *discontinuous* process governed by qualitative transformations resulting from sequential gene derepression [*Flickinger*, 1962; *Denis*, 1966], the substances specific for a given tissue appearing only at the time of terminal differentiation (fig. 6 b). Nevertheless, the fact that increasing numbers of genes are being repressed during development has been clearly demonstrated. The mRNA of embryonic cells is of higher sequence complexity than that of cells that have attained their definitive

---

Fig. 6. Comparison between the two concepts of tissue differentiation. Above: The novel concept. The level of differentiation rises progressively as the autonomous progressions succeed each other. At the same time a dominant luxury metabolic strategy is progressively selected (histogram). When this exceeds a certain threshold level (T) terminal differentiation takes place. Successive determinative events re-start autonomous progressions. Below: The classical concept. Each determinative event activates new genes. The gene that codes for the luxury molecule specific for the tissue in question is only derepressed when terminal differentiation takes place (histogram). The level of differentiation increases stepwise.

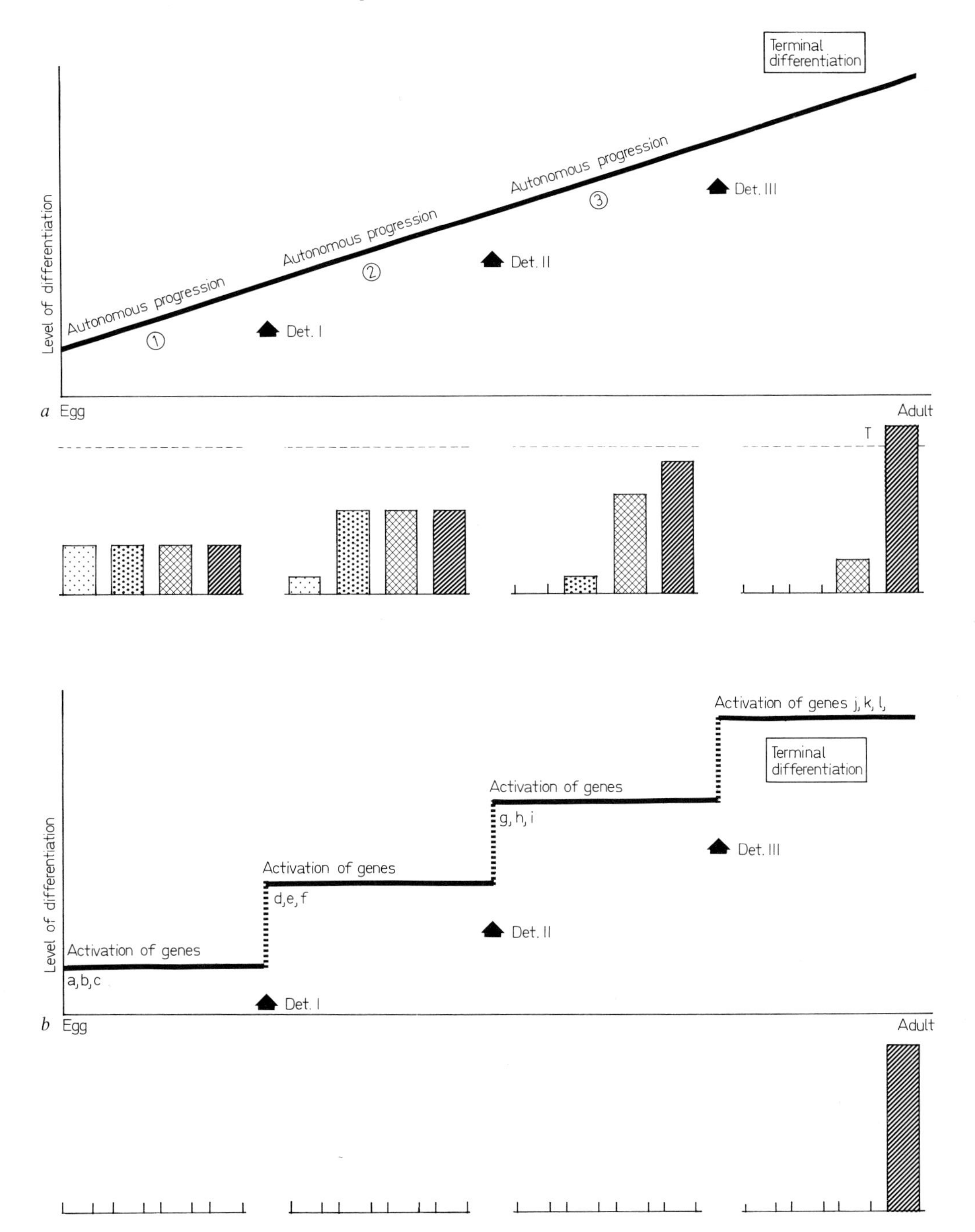
Terminal differentiation
Level of differentiation
Autonomous progression
①
Det. I
Autonomous progression
②
Det. II
Autonomous progression
③
Det. III
a Egg
Adult
T
Activation of genes j, k, l,
Terminal differentiation
Activation of genes
g, h, i
Det. III
Activation of genes
d,e, f
Det. II
Activation of genes
a,b,c
Det. I
b Egg
Adult

differentiation [*Frenster and Herstein,* 1973]. The same holds for dedifferentiated cells that are redifferentiating, particularly for regenerating liver [*Church and MacCarthy,* 1967] and for Friend leukaemic cells as they differentiate into erythrocytes, producing more and more globin [*Affara and Daubas,* 1979]. Other quantitative changes during tissue differentiation are known with respect to the composition of protein families (for instance crystallins during lens differentiation, [*Clayton* et al., 1976; *MacAvoy,* 1980] and 'fetal' and 'adult' haemoglobins during erythrocyte differentiation [*Zagris and Melton,* 1978]).

On the basis of numerous observations of this kind *Spiegel* [1960] and later *Caplan and Ordahl* [1978] have already presented theories based on sequential gene repression. These must have appeared illogical to most people, as would the idea that a child would know everything on entering school but would have forgotten almost all of it on leaving the university. In actual fact, what is important in the differentiation of a tissue is the intensification of one luxury metabolic strategy [*Clayton,* in preparation] at the expense of all the other strategies; the latter can be maintained at a low level or disappear without affecting the normal function of the tissue, the global amount of 'cell work' being done remaining the same.

The specific activities of a tissue are stimulated in the course of successive autonomous progressions made possible by homotypic cell interactions. The realisation of this fact, long after the discovery of embryonic induction, has failed to change the views of theoretical embryologists. They still think the only possibility for the cells to attain their terminal differentiation is by way of heterotypic actions, each of them taking the cell lineage from one 'station' to the next by activating new genes, whose role in the final establishment of tissue activities is difficult to explain. This view has led to a mistaken idea of the dynamics of differentiation (fig. 6) by assuming that inductions and hormonal influences represent the prime movers of development. In reality, however, *each heterotypic action only interferes with an autonomous progression that is already going on. It would remain without effect if the homotypic interactions would not establish the necessary state of competence* (particularly in the case of early embryonic inductions) *and then effectuate the previous determination of the induced tissue.* A heterotypic action does not necessarily always re-launch an autonomous progression that has run out. It may also deviate it from the course it had taken, or stop it temporarily. But, most important, differentiation may sometimes continue in the same manner in the absence as in the presence of a heterotypic action. In other words, *it is autonomous progression itself that appears to represent the prime mover of development.* If that is so, the registration of extracellular information is comparable to a sort of 'remote control' for autonomous progression. In fact, the instructions provided by a remote control system to a machine in motion do not propel the machine but control its movements (starting, stopping and change of direction), and a fault in the remote control does not ipso facto result in the immediate stoppage of the machine.

The diversification of the cells involved in the production of a single tissue occurs on similar principles. During their autonomous progression the cells must continuously adjust their activities to the changes occurring in the individualities of neighbouring cells, whether these belong to the same or to a different cell population. Because they did not realise the importance of autonomous progression theoretical embryologists, in order to explain the structuring of tissues, have had to take recourse to the idea of supracellular or global control: either by diffusion of morphogens [*Wolpert,* 1969; 1971] or by other morphogenetic signals [*Goodwin and Cohen,* 1969]. This global control would be exerted continuously throughout the animal's life.

Each time a heterotypic action occurs, extracellular information is registered by the cells and subsequently interpreted in terms of the cytoplasmic information that is already present, which is in its turn modified. Therefore, *each heterotypic action adds its effects to those of the previous one.* By way of example, let us consider a cell of a urinary tubule in the metanephros of an adult amphibian. Before becoming functional under the influence of thyroxine it formed part of a cell aggregate induced in the nephrogenic mesenchyme by the pronephric duct. This mesenchyme in its turn originated by dissociation of part of the intermediate mesoderm determined in the lateral mesoderm under the influence of the archenteron roof. The mesoderm itself had been formed under the inductive influence of the yolk endoderm of the blastula.

The progression of differentiation may be usefully compared to the working of a computer into which at intervals are fed data representing the outputs of other computers. Its memory is progressively enriched while its calculations become more complicated. The ultimate result depends just as much on the information content of the computer memory as on the order in which new data are fed into it. By analogy, *all features of the individuality of a cell in the organism are rigorously determined by the summation of the different kinds of extracellular information registered by its predecessors in the line of ascent.* This computer analogy should not make one think that differentiation comes to a stop because the cytoplasmic memory becomes 'saturated' and can no longer register information. On the contrary, the maintenance of cellular individualities requires the constant exchange of positional information – and in some cases the circulation of hormones. When the cells are isolated their 'collective memory' is completely erased, which is shown most clearly in cases of dedifferentiation followed by transdifferentiation. Whichever cell lineage one considers, it is always the registration of extracellular information that enables it to initiate and then to maintain specific activities that it would be unable to sustain by itself. In this regard the extracellular information is comparable to a crick that lifts an object and sustains it at a higher level.

It is now possible to understand why during the entire life of the organism the successive individualities that the cells come to possess rigorously depend on their position in the whole. Most of the information registered by the line of ascent is positional information. *The individuality of an adult cell therefore depends on the position the various cell generations in its 'ascendant line' have occupied in the course of development.* This is what one may call its 'positional history', to use a term that *Wolpert* [1969] coined without, however, giving it a precise meaning. As to the information that allows the cell to maintain its individuality in the adult organism, this depends on the individualities of the neighbouring cells that provide the information, and these individualities in their turn were acquired by the same process of summation of positional information effects. In other words, the cells simply 'help' each other to maintain the individualities they have acquired by the same sort of 'mutual aid'. To explain position effects it is theoretically unnecessary to take recourse to the action of supracellular controls whose existence has not been demonstrated. Some modification in the neighbourhood of a cell lineage (either in information content or in the sequence of information registered) has repercussions for the differentiation of the lineage in question, and more particularly on the individuality of the cell representing the end of the lineage. Let us return to the urinary tubule cell. If one of its ancestors had not been trapped in a cell aggregate (particularly because the pronephric duct was not present in its vicinity) it would have become a connective tissue cell. If the cells forming the intermediate mesoderm had been closer to or farther away from the archenteron roof they would have formed products of the somites or the lateral plate. If the blastomeres forming the lateral mesoderm had been located further away from the yolk endoderm they would have formed ectoderm and perhaps, after contact with the archenteron roof, neural plate cells.

The principle of the summation of positional information seems to be put in default by the fact that it is possible to replace a natural inducer by various other tissues. However, we should not forget that induction is a special case of the general class of heterotypic actions. The registration of the information supplied by the inducer only effects a choice from among a small number of autonomous progressions of equal probability (usually two). But cell individuality has other features whose determination does seem to require a rigorously correct summation of information. For example, the determination of a placode requires the cooperation of several neighbouring tissues which all act in the same manner. Although the absence of one or even several of these tissues does not necessarily lead to a change in the sort of placode produced, it nevertheless prevents its normal differentiation [*Jacobson*, 1966].

The best model one can propose to account for the automatism of differentiation is that of a goal-seeking missile. The trajectory of the missile, from the launching-pad to the goal, represents the transformation that a cell lineage ondergoes from the egg to the acquisition of its definitive individuality. The trajectory is not fixed from the start. The firings (by radio signal) of the successive stages of the rocket, which each propel the missile in a given direction, represent the various determinative events that initiate the successive autonomous progressions and determine their particular features. The corrections of the trajectory as a function of objects detected on the way, performed by the computer in the head of the missile, represent the readjustments of the cells engaged in autonomous progression. If upon firing of a given stage of the rocket the missile receives no further information it will for a time continue its course in a direction defined by the interpretation of the information it had received previously; this represents the behaviour of embryonic cells removed from all heterotypic action.

To conclude, then, all understanding of the dynamics of differentiation rests on the comprehension of homotypic interactions. Unfortunately we know next to nothing about the way in which cell-to-cell contacts control gene activity and are involved in the selection of luxury metabolic strategies. This is clearly an important task for the future.

The Reproduction of Structural Patterns

Structural patterns are so enormously diverse that just the external appearance of any remaining dead part of an animal enables one to identify it with certainty, that is, to determine both where it belongs in the body and where the organism belongs in the classification of animals. This is not to say that a structural pattern is entirely composed of unique features. Because most of them can also be found elsewhere, albeit in many different variants, we can consider them as pattern 'motifs'. Certain motifs are not even unique to animals but can also be found in plants and, even more remarkably, in colonies and aggregates of protozoans, those primitive pluricellular forms the knowledge of which considerably facilitates the analysis of structural patterns in animals.

The most fundamental characteristic of any pluricellular structure is 'polarity'. The organism has at least two dissimilar poles, ends or surfaces. For instance, a young colony of the colonial flagellate *Volvox* is spherical in shape and consists of a single layer of identical cells; an opening on one side defines an axis of symmetry that is at the same time the axis of polarity. In motile forms the direction of movement coincides with the axis of polarity, which here is also called the antero-posterior axis. We mention as an example the 'slug' of the cellular slime mould *Dictyostelium*, which originates from an aggregate of originally free-living soil amoebae and later differentiates into a fruiting body. While the slug is moving about it does so always with the same end forward: the end that contains the cells that were the first to aggregate. In true animals, some lower groups excepted, the surface that is 'ventral' during movement shows a different organisation from that of the opposite 'dorsal' surface. Thus, the antero-posterior polarity is combined with a dorso-ventral polarity, so that axial symmetry is replaced by bilateral symmetry. Polarity manifests itself at all levels along the polar axis, even though they may appear identical and devoid of special structure. A good example of this is the fresh-water polyp *Hydra*. Here polarity is morphologically defined by the presence of a mouth and a ring of tentacles at one end and an adhesive 'foot' at the other end of the otherwise unstructured body column. When the latter is cut into pieces these regenerate, each of them retaining its polarity: the tentacles always appear on the cut surface that was located closest to the original oral end. This property of polarity is found even in the most highly evolved animals. Although there a multiplicity of features makes the patterns more complicated, the organism as a whole possesses an antero-posterior and a dorso-ventral polarity, which can both be demonstrated to be present in all organs. If organs are paired they often also have a medio-lateral (or a proximo-distal) polarity.

Structural patterns are characterised by a universal motif called 'graded pattern'. By this we mean a gradual variation of certain features of cell individuality along an axis; this is usually one of the polar axes of the organism or organ but can also have a quite different direction. In general, in both embryonic and adult cell populations it is the invisible features of cell individuality which are involved in the formation of a graded pattern; this can only be brought to light by experiment (p. 10). Certain covert graded patterns are bound up with variations in the metabolic standing of the cells. They then manifest themselves in what has been called a 'metabolic gradient' [*Child*, 1941], for instance a graded variation in oxygen consumption, or in the activity or concentration of a particular substance. Metabolic gradients are ubiquitous: wherever one has looked for them one has found them, in the most varied types of organisation and in embryos as well as adults [*Child*, 1941]. The visible features of cell individuality define an overt pattern that looks quite different from the covert graded pattern. For example, in planarians stained vitally under anaerobic conditions the cells are discoloured more rapidly as they are located more closely to the anterior end (fig. 7 a). The visible organ pattern of the animal in no way reflects the presence of this gradient. Other covert graded patterns are bound up with variations in cell adhesiveness. Such a graded pattern has been shown to exist in a butterfly pupal wing, where again it has nothing whatever to do with the overt pattern of the wing [*Nardi and Kafatos*, 1976]. Small pieces of tissue taken from various proximo-distal levels of the pupal wing and grafted into other levels behave in different manners, such as contraction or expansion of the graft, changes in the density of scales upon differentiation, rosette formation or inversion of polarity, because the cells arrange themselves in different ways according to the site of origin and the axial distance of transplantation.

Covert graded patterns play a role of paramount importance in morphogenesis, in that cells that are going to transform themselves do not do so all at the same time (e.g. differentiation starts at one end of the cell population and gradually proceeds to the other end) or do not all develop in the same manner (e.g. various different regions may form without any information

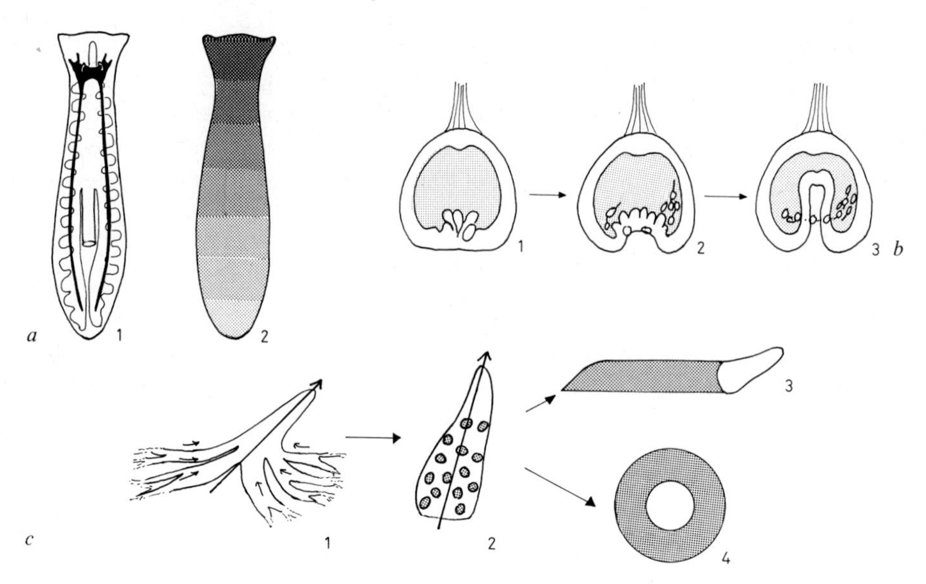

Fig. 7. Overt and covert structural patterns. *a* Planarians: 1, overt pattern; 2, covert graded pattern revealed by measurements of oxygen consumption. *b* In the sea urchin blastula cell adhesiveness diminishes from the animal to the vegetal pole (1). As a result micromeres dissociate themselves at the pre-gastrula stage (primary mesenchyme cells) (2), while the archenteron invaginates at the gastrula stage (3). *c* In *Dictyostelium* during aggregation an antero-posterior polarity (1) is established. The prespore cells (dotted) are at first uniformly distributed in the slug (2). Later they regroup themselves in the hind part of the slug, at the end opposite to what was earlier the aggregation centre (3). Upon disaggregation of a slug the prestalk cells, which adhere strongly to each other, reaggregate in the centre, while the prespore cells, which are less adhesive, end up in the periphery (4).

coming from outside the cell population). In more precise terms, *the presence of a covert graded pattern may manifest itself in later phases of development in the specific features of a spatio-temporal pattern.* We will mention three simple examples. In a young *Volvox* colony the cells are apparently identical. In the course of time they change in shape and become prismatic, their inner surface being larger. This change starts at the pole opposite to the anterior opening and extends to the whole colony, as a result of which it everts itself. In the blastula of the sea urchin the specific topography of the gastrula is foreshadowed by a graded pattern in cell adhesiveness, which diminishes from the animal to the vegetal pole [*Gustafson,* 1965]. The cells of the animal pole area, which bear the apical tuft of cilia, closely adhere to one another; those of the vegetal pole area dissociate and form the primary mesenchyme inside the blastocoele. The cells a little higher up the gradient, which now come to occupy the vegetal pole area, only loosen their contacts on the blastocoelic side, as a result of which the blastula wall alters its shape and invaginates as the archenteron (fig. 7b). The morphogenesis of mammalian tooth buds is due to an epidermal-mesenchymal interaction. The size of the buds decreases from back to front along the maxilla, apparently because due to the existence of a graded pattern they are formed from progressively older cells [*Lumsden*, 1979].

Various other pattern motifs do not have the universality of graded patterns but nevertheless are found at all levels of animal organisation, usually in specific tissue types: acini, tubules, cell condensations, hexagonal patterns, etc. The various sorts of common pattern motifs necessarily reflect common cell properties or activities. Before we start the analysis of spatio-temporal patterns we must first understand how they arise.

The Common Motifs of Structural Patterns, and How They Are Determined

> The emergence of graded patterns is due to the diversification of positional information as a result of mitosis, and to the interference of various processes (such as propagation of information, cell aggregation, apical growth and tissue renewal) with the autonomous progression of differentiation. Other common motifs of structural patterns are determined independently of graded patterns but also result from elementary social cell behaviour.

At the end of the second chapter (p. 23) we formulated the principle that two cells which possess the same cytoplasmic information can only remain identical if they register the same positional information. This requires that their environments are absolutely identical: the same number of cells with the same individualities. This requirement meanwhile does not carry its full force for those visible features of cell individuality whose realisation is a question of all or none (such as specific tissue structures). It is most stringent for the mostly invisible features bound up with the metabolic standing of cells, which can take all values comprised between two extremes.

If we accept this principle it becomes clear that the acquisition of the pluricellular state automatically implies the emergence of graded patterns –

most of them invisible. For this emergence two models have been proposed earlier [*Chandebois,* 1977], both for cells which after mitosis remain arranged in a single file. The first model (fig. 8 a) starts from two cells with different individualities A and B (one of them could for instance contain a metabolic inhibitor which is absent in the other). When both cells have divided once their daughter cells retain the cytoplasmic information of the mother cell. However, the cells at the two ends of the row only receive information from their sister cell, while those in the middle both receive information from an A cell on one side and a B cell on the other. Once this information is registered the four cells have acquired four different individualities: AAO, AAB, BAB and BBO. Again, after the second division the two daughter cells of each cell do not register the same positional information. For instance, one of the AAO cells only receives information from the other AAO cell, while the latter gets information from AAO on one side and from AAB on the other. One of the AAB cells gets information from AAO and AAB, the other from AAB and BAB. As this process repeats itself at every division ultimately no two cells have exactly the same individuality. If one calculates the ratio A/B for all cells, for instance just prior to the third division, one finds that it decreases from one end of the row to the other as follows: 4/0, 6/1, 6/2, 5/4, 4/5, 2/6, 1/6, 0/4. The closer a cell is to one end, the more it resembles the cell that occupies that end. Therefore, as soon as we have two different cells the sequence of divisons *automatically* engenders a structural pattern: this is what we call the 'asymmetrical graded pattern of growth'. In the second model (fig. 8b) the first two cells are identical. Of each of the two cell pairs produced by the first division one daughter cell is at the end of the row while the other is flanked by two cells. They therefore do not register exactly the same positional information and take on different individualities, A and B. For the descendants of each of these two pairs the process of exchange of information is the same as in the first model, the result being a 'symmetrical graded pattern of growth'.

During the autonomous progressions which succeed each other throughout development the individualities of the cells are gradually transformed.

Fig. 8. Models for the emergence of graded patterns. *a* Asymmetrical graded pattern of growth. *b* Symmetrical graded pattern of growth. *c* Graded pattern of accretion (the cells added to the system are produced by generative cells, G). *d* Graded pattern of accretion (the cells are added to the system by an aggregation process). *e* Graded pattern of induction (I = inducer; S_1-S_4 = successive steps of the autonomous progression). *f* Graded pattern of cell renewal (G = generative cell; † = dying cell). Explanation in text.

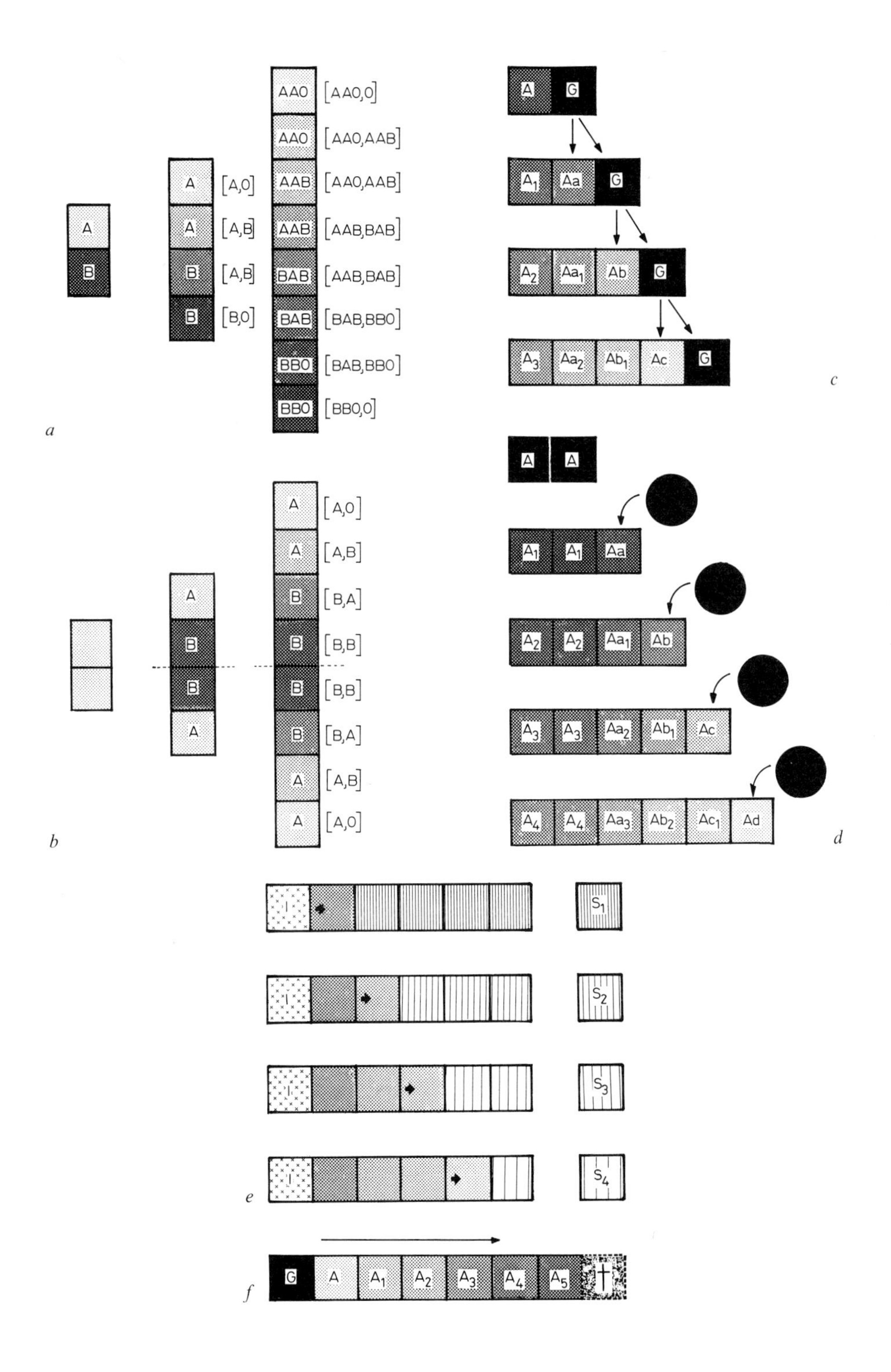

A
B
A
A
B
B
[A,0]
[A,B]
[A,B]
[B,0]
AA0
AA0
AAB
AAB
BAB
BAB
BB0
BB0
[AA0,0]
[AA0,AAB]
[AA0,AAB]
[AAB,BAB]
[AAB,BAB]
[BAB,BB0]
[BAB,BB0]
[BB0,0]
a
A
G
A1
Aa
G
A2
Aa1
Ab
G
A3
Aa2
Ab1
Ac
G
c
A
B
B
A
A
A
B
B
B
B
A
A
[A,0]
[A,B]
[B,A]
[B,B]
[B,B]
[B,A]
[A,B]
[A,0]
b
A
A
A1
A1
Aa
A2
A2
Aa1
Ab
A3
A3
Aa2
Ab1
Ac
A4
A4
Aa3
Ab2
Ac1
Ad
d
I
S1
I
S2
I
S3
I
S4
e
G
A
A1
A2
A3
A4
A5
f

However, various cellular activities interfere with the course of the autonomous progressions, from which again the emergence of graded patterns automatically results. Several models for simple linear arrays of cells have been presented earlier [*Chandebois,* 1977].

If a cell in a cell population engaged in autonomous progression is made to change its individuality in response to some signal, the neighbouring cells will readjust their own individuality, followed by cells located further away, and so on. Because the changes will affect cells that are progressively more advanced in their autonomous progression a graded pattern will necessarily ensue: such a pattern we call a 'graded pattern of induction'. When for example an inductive effect propagates through a cell population it will reach cells that are closer and closer to the time of loss of competence (fig. 8e). Below we will encounter an example of this concerning the organisation of the archenteron roof. The characteristics of such a gradient are obviously not determined by the inducer but only by the rate of cell transformation and of transmission of the induction. In other words, the graded pattern is established automatically and is determined by the programme of the autonomous progression.

We have seen that cells which are aggregating are thereby determined to enter a new autonomous progression. If the aggregate grows by accretion of cells at its periphery a graded pattern necessarily results. For instance, in the precartilage condensations of the developing limb, cell shape and stainability vary gradually from the centre to the periphery [*Ede and Flint,* 1972]. Each time a cell joins the aggregate it will start its new autonomous progression with a complement of cytoplasmic information that is different from that of the cells already in the aggregate (fig. 8d). Let us start from two A cells which have just aggregated and are embarking on an autonomous progression A_1, A_2, A_3, etc. When the cells have reached stage A_1 a third cell joins the aggregate. The information coming from the A_1 cells will force it to engage in a different autonomous progression A_a, A_{a1}, A_{a2}, etc. It will then communicate this information to the fourth cell, which will start on an autonomous progression that is different again: A_b, A_{b1}, A_{b2}, etc. Again, the establishment of this 'graded pattern of accretion' is an automatic process. Its characteristics are determined partly by the rate of aggregation and partly by the rate at which the cells change their individualities. Its programme is therefore again that of the autonomous progression.

In some cell populations, when they reach a certain degree of differentiation, mitotic activity may be restricted to, or at least be much higher in a particular region. The cells in this region must be kept in an undifferentiated state, which requires information coming from a different adjoining cell population (p. 47–48). In the most simple theoretical case one of the daughter cells of a dividing generative cell escapes from the range of influence of this information and embarks upon its autonomous progression A_1, A_2, A_3, etc. Because every new cell that is produced 'pushes' the cells already present away from the proliferation zone, the closer a cell is to the other end of the row the more advanced is it in its progression (fig. 8c). In addition, each time a new cell enters the system it starts its autonomous progression with a different content of positional information than the previous cells – just as in the case of aggregation. In the embryo this sort of graded pattern is established in certain organs which show localised proliferation at the apex. An example is the limb bud, which grows apically thanks to a proliferation zone, which is maintained by the apical ridge that is formed by the overlying epidermis. The proximo-distal polarity of the limb is bound up with this apical proliferation. The mesodermal cells are progressively more advanced in their differentiation as they are located further away from the tip. This has been demonstrated by measuring the concentration of two different proteoglycans: hyaluronate and chondroitin sulphate, the former being predominant in undifferentiated mesoderm and the latter in cartilage. The concentration of hyaluronate is maximal close to the epidermal ridge and declines gradually and regularly in a proximal direction; the converse situation holds for chondroitin sulphate [*Kosher* et al., 1981]. The features of this type of 'graded pattern of accretion' are determined by the time scale of two different automatic processes, which are both programmed in the cells of the proliferation zone: (1) the rate of their autonomous progression, determined by the cytoplasmic information stored previously, and (2) their mitotic frequency, determined by the positional information they are registering (which in the example of the limb comes from the apical ridge). In adult animals this type of graded pattern can be maintained by cell renewal (fig. 8f). A fairly simple example is provided by the fresh-water polyp *Hydra*, where cells are continuously produced in the body column (though not in a localised growth zone), to die and be sloughed off both at the tips of the tentacles and in the foot region. The resulting 'double' graded pattern defines the apico-basal polarity of the organism. This polarity can be reversed experimentally but this requires several days and involves dedifferentiation and cell death [*Campbell*, 1979].

There are other, more varied, kinds of pattern motif which are not bound up with graded patterns but are nevertheless also determined by changes in the elementary social behaviour of the cells. In certain cases such changes are correlated with an autonomous progression (we have already described cases involving cell adhesiveness or short-range inhibition, p. 29–30). In a structural pattern any one motif necessarily occurs in combination with various other visible traits and with one or more covert graded patterns.

In a human society a man belongs simultaneously to several groups: a family, an association, a club, etc. In each of those he is engaged with a particular aspect of his personality. Moreover, these various kinds of group were formed at different times during the history of the society, each with its specific motivation. The structuring of pluricellular organisms is based on altogether comparable principles. One and the same cell is involved in various motifs or traits of the pattern of the population to which it belongs, for each of them with a particular feature of its individuality. For example, it can form part of a covert graded pattern because of its metabolic standing, and at the same time form part of an acinus as a result of its cell surface properties. The various motifs and features of the structural pattern are not all of the same 'age' and have been determined on the basis of information of different origin. This is very well illustrated by experiments made on *Dictyostelium* (fig. 7c). In the slug prestalk and prespore cells are randomly distributed except in the anterior tip, which consists entirely of specific prestalk cells: those which were the first to aggregate [*Müller and Hohl,* 1973]. The transformation of the slug into a fruiting body represents an instance of 'sorting out'. When a slug is dissociated the prespore cells, which adhere less strongly to each other, rearrange themselves at the periphery of the reaggregate [*Garrod and Foreman,* 1977]. In the intact slug this process interferes with the antero-posterior polarity obviously acquired during aggregation (i.e. with a covert graded pattern of accretion). As a result the prestalk cells rearrange themselves in the front part of the slug, joining those which had formed the aggregation centre.

Summarising, *when the various features of a structural pattern are determined at the same time they 'interfere'* (although in English the verb 'to interfere' in this sense is now chiefly used in physics, e.g. 'interfering waves', we use it in the original extended sense, where it may apply to all phenomena, physical and non-physical; our use does not imply that physical phenomena per se are involved). *This is because they result from different activities of the same group of cells. When they are determined successively one is so to speak superimposed on the other. Thus, the structural pattern complicates itself by summation.*

Throughout animal development exchange of information occurs between the various cell populations. These heterotypic actions interfere with autonomous progressions and thus determine new features which complicate the features established by the elementary social behaviour of the cells. We will give here a typical example. In the mammalian salivary gland the typical branching of the epithelial component does not take place when it is isolated from the mesenchymal component. It does occur when the two are recombined immediately upon separation, but the mesenchyme cultured alone for 20 days loses its capacity to support the branching morphogenesis of freshly isolated epithelium. This change in inducing properties is accompanied by a loss of cell adhesiveness in the mesenchyme [*Grobstein*, 1953].

It is clear, therefore, that tissue specialisation in the various animal forms brings with it a considerable increase in complexity of structural patterns, which makes it particularly difficult to analyse experimentally how they are reproduced during ontogenesis.

Interference of Pattern Features

The unfolding of a spatio-temporal pattern in a cell population in vivo results from the readjustment that the population as a whole must perform during its autonomous progression. This readjustment is an entirely automatic process, to which corresponds a specific developmental programme. In the latter one can distinguish various elements which we call 'prepatterns'. Each prepattern is established using a separate source of information and imparts certain specific features to the definitive pattern. The information already stored in the collective memory of the cell population at the start of its autonomous progression establishes two sorts of intrinsic prepattern: the elementary social prepattern (the programme of the autonomous progression) and the antecedent prepattern (the organisation the population had acquired previously). During the autonomous progression other, extrinsic prepatterns are provided by neighbouring cell populations. The latter may act (1) through the individualities of their cells (environmental prepattern), (2) through their spatial organisation (imprinting prepattern), or (3) through their relative position (positional prepattern). The features imparted by various prepatterns during a readjustment interfere with each other.

The analysis of the reproduction of a structural pattern requires the unambiguous definition of a unit of time and space. The best convention is to follow a homogeneous cell population from the time it is launched on a progression that can continue autonomously in vitro, up to the time a visible pattern is established.

According to the principles we have expounded earlier, when cells engaged in the same autonomous progression do not all start with the same cytoplasmic information content, or when certain cells come under the influence of heterotypic actions, each cell must constantly readjust its individuality to the changes occurring in its neighbours. In effect the cell population as a whole itself performs a 'readjustment' [*Chandebois,* 1976a, b]. Simple theoretical models suffice to understand this notion (fig. 9). Let us consider a cell population which has just been determined. It is entirely homogeneous and would remain so if allowed to effect its autonomous progression in vitro (fig. 9 a). If it is surrounded in vivo by the cells of another homogeneous population (fig. 9 b) the influence of the latter will propagate through it. Concentric layers of cells will appear whose ultimate cell individualities will differ because the heterotypic influence has reached them at progressively later stages in their autonomous progression. Let us now assume (fig. 9 c) that this same population is surrounded by two different homogeneous populations. The heterotypic influences propagate in the same manner as before, but now in the central population there appear two different regions under the influence of the two other populations. Since these regions will also influence each other, the number of cell types increases by much more than a factor of two. On either side of the boundary between the two regions the cells will take on particular individualities and will also induce their neighbours to do the same. Obviously a readjustment is accompanied by the unfolding of a spatio-temporal pattern; the features of its organisation and the time schedule of their appearance are exclusively determined by the individualities of the cells present at the beginning.

This scheme is of course an oversimplification. Other phenomena will yet accentuate the diversification of cell individualities: particularly cell divisions leading automatically to the establishment of visible or invisible graded patterns, and cell displacements giving rise to new exchange of information. In addition, the increase in complexity of a structural pattern is not only related to the diversification of cell morphology. Changes in cell properties contribute even more to the ultimate form of the pattern: changes in cell adhesiveness resulting in specific pattern features such as acini or tubules,

---

Fig. 9. Scheme of the readjustment of a cell population surrounded by one *(b)* or two *(c)* other homogeneous cell populations, compared with autonomous progression in vitro *(a)*. Explanation in text.

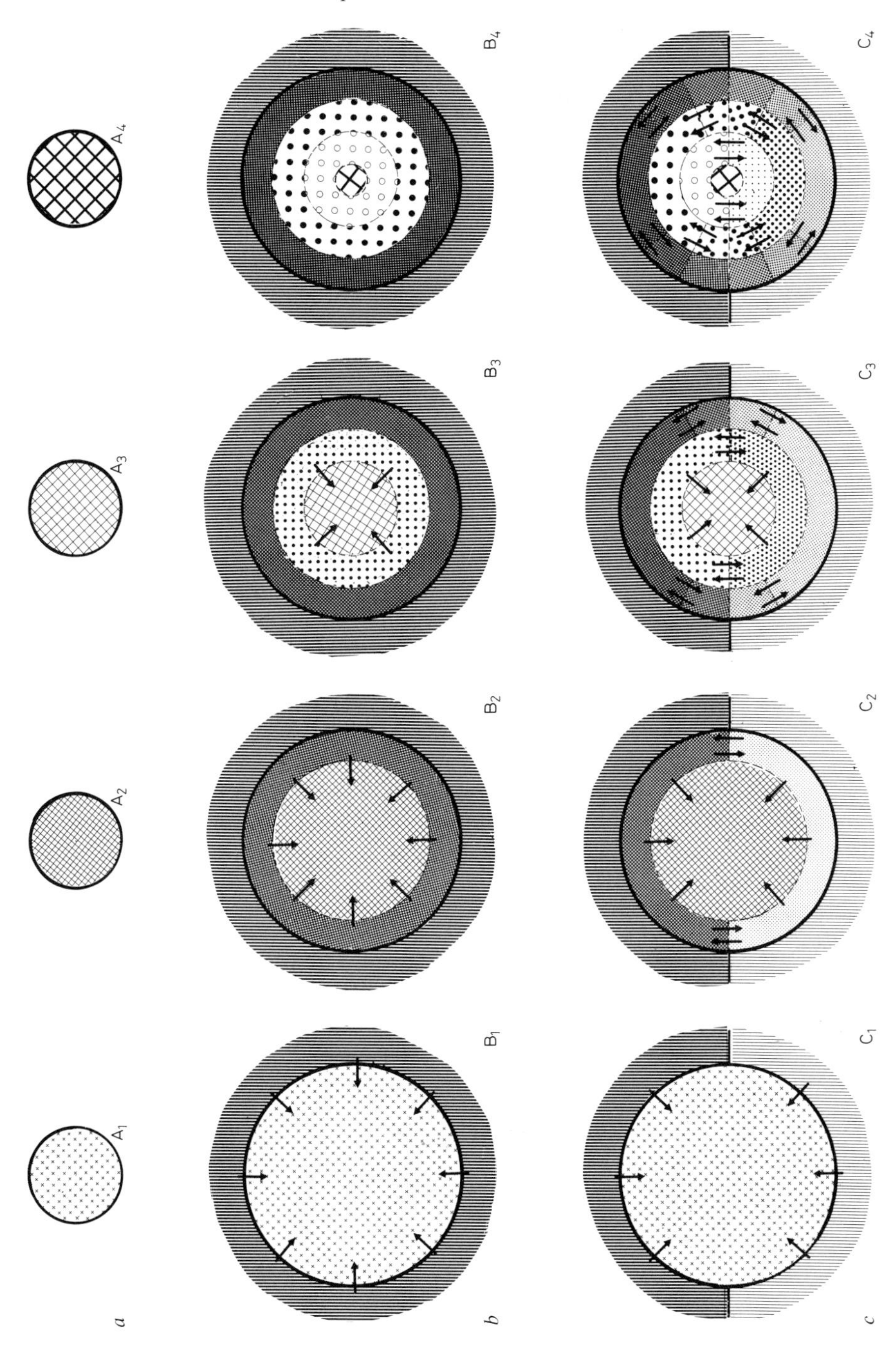
a
A1
A2
A3
A4
b
B1
B2
B3
B4
c
C1
C2
C3
C4

and differential growth and cell death which may distort other features of the pattern. Finally, at least one of the starting cell populations may have acquired a certain invisible organisation prior to the readjustment, which will give rise to even greater complexity. It is theoretically plausible that at the end of the autonomous progression of a given cell population no two cells will be strictly identical any more.

The unfolding of a given spatio-temporal pattern in a homogeneous cell population in vivo *is an entirely automatic process.* It represents the execution of a specific programme which we have in an earlier publication called the 'developmental programme' of the cell population: it is a programme of cell interactions [*Chandebois,* 1980a]. It is important to realise that at the time of effective determination the cell population is not yet in possession of the whole programme: this is because throughout the readjustment of the population its cells continue registering information coming from other, neighbouring cell populations. *One may therefore distinguish in the developmental programme a number of components which are established using diverse, experimentally dissociable sources of information. Each of these components imparts certain features of the ultimate pattern, which is why they may be called prepatterns* [*Chandebois,* 1977].

Those prepatterns that are already present in the cell population at the start of its readjustment we call 'intrinsic prepatterns' (fig. 10). The organisational features which are due to the elementary social behaviour of the cells are fixed by the programme selected at the time of determination of the cell population; this is the only programme that is executed during autonomous progression in vitro following disaggregation. Thus, one can say that the programme of the autonomous progression brings with it an 'elementary social prepattern'. In certain cases disaggregation prevents the appearance of certain features which would appear without disaggregation. In those cases the cell population had obviously acquired a certain invisible organisation prior to its determination. This organisation serves as an 'antecedent prepattern'.

Those prepatterns which derive from other cell populations during the readjustment we call 'extrinsic prepatterns' (fig. 10). Three kinds may be distinguished. Certain features are determined simply by the presence of an adjoining different cell population (propagation of the heterotypic action or modification of elementary social behaviour); the information originating from the latter population brings with it an 'environmental prepattern'. When, on the other hand, more than one other cell population is present near the one effecting its readjustment, certain features depend on their relative position: they provide a 'positional prepattern'; in this case, if the transform-

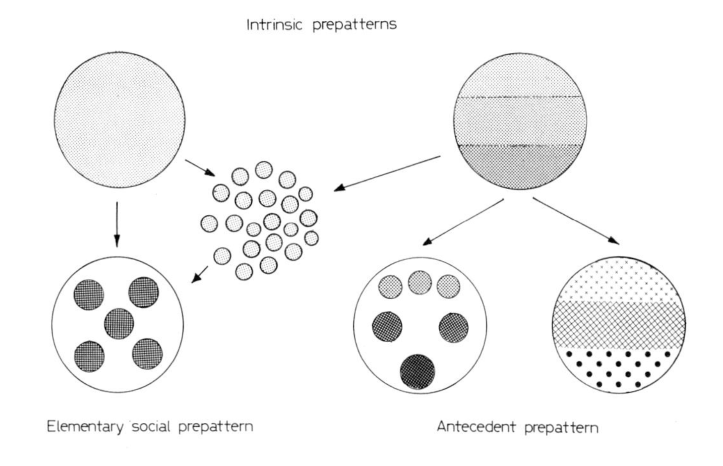

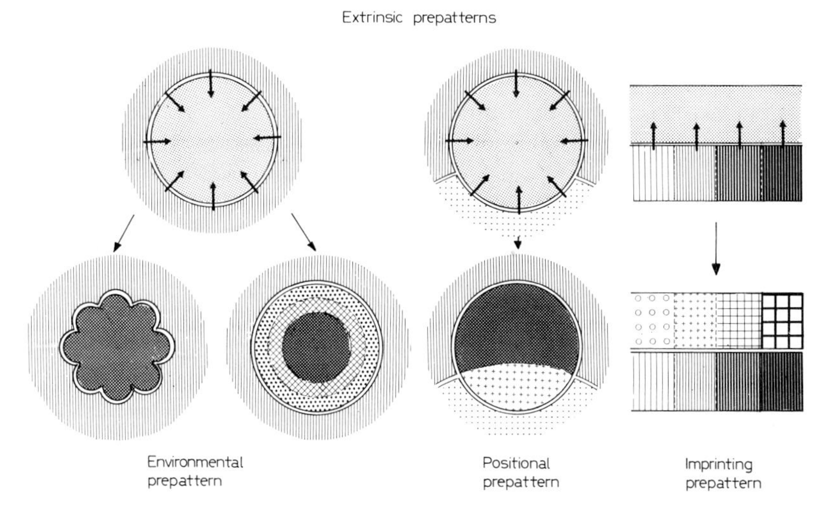

Fig. 10. Various classes of prepattern. Explanation in text.

ing cell population already possesses a certain organisation a simple rotation may result in a considerable alteration of the definitive pattern. Finally, if an adjoining cell population has itself a visible or invisible organisation it will impart to the transforming population certain features analogous to its own incipient pattern and thus will serve as an 'imprinting prepattern'. Given the fact that one cell registers and interprets information of different origins, the effects of the various prepatterns are not sharply separated spatially: they interfere with each other.

This scheme of the reproduction of a structural pattern is largely based on the best known aspects of vertebrate organogenesis. We will recall these in the succeeding sections not only to justify the scheme and to provide examples of the application of its terminology. We also want to show that each readjustment occurring in a given rudiment at a particular stage is a *unique* event in the embryo, even though it is based on universal principles. The special features of each readjustment must be established by exhaustive experimental analysis; they cannot be arrived at by any theoretical attempt whatever.

The Organisation of the Mesoderm in Amphibians

The mesodermal mantle of the gastrula originates from 'animal' blastomeres which were originally destined to give rise to ectoderm, but which at the blastula stage have undergone a deviation of their autonomous progression under the influence of the large cells of the yolk endoderm in the vegetal half of the embryo [*Nieuwkoop,* 1969a]. This influence extends from cell to cell in the direction of the animal pole. The process probably stops when the cells lose their competence to respond [*Kurihara and Sasaki,* 1981], as a result of which the future ectoderm becomes restricted to the animal region of the blastula.

Starting with the neurula stage the organisation of the mesoderm is already quite complex. The archenteron roof will form pharyngeal endoderm, prechordal plate and notochord from front to back. The lateral mesoderm will form the somites, the intermediate mesoderm and the lateral plates, which when isolated in vitro will differentiate respectively into striated muscle, kidney and blood. As a result of the progressive character of the induction exerted by the yolk endoderm, differentiation in the mesodermal mantle proceeds from front to back (it should be remembered that during gastrulation the orientation of the mesoderm becomes inverted).

Although in normal development the grey crescent represents the prospective notochord, it has nothing to do with its determination. If one recombines an isolated animal half of a blastula with an endodermal half in arbitrary orientation (e.g. after rotation through 90°), with their dorsal sides marked in some manner, one finds that the position of the embryonic axis is always determined by the endoderm [*Nieuwkoop,* 1969b].

Spemann's [1931] experiment involving the transplantation of the dorsal blastoporal lip (p. 31) suggests that the appearance of somites and intermediate mesoderm in the host lateral plate mesoderm is due to the inductive effect

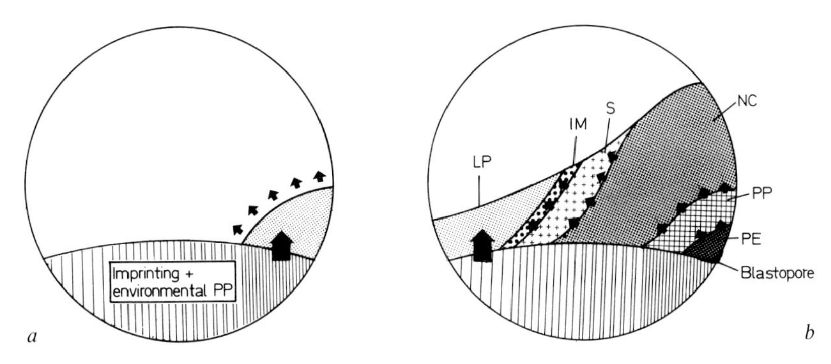

Fig. 11. Interference of prepattern effects in mesoderm induction. *a* Start of induction on the dorsal side. *b* End of the inductive period (induced structures projected back onto the gastrula). im, Intermediate mesoderm; lp, lateral plate mesoderm; nc, notochord; pe, pharyngeal endoderm; pp, prechordal plate; s, somitic mesoderm; large arrows, induction by the vegetal yolk mass; small arrows, propagation of induction.

of the archenteron roof. In vitro culture of the various prospective areas of the mesoderm with or without prospective notochord [*Yamada,* 1940] has shown that to form somites the cells must first be 'elevated' to the level of intermediate mesoderm and then remain under the influence of the notochord while they are still competent. However, the notochord only transmits the induction such as it has itself received it. In effect the lateral portion of the yolk endoderm, if associated with ectoderm possessing mesodermal competence, induces somites and intermediate mesoderm [*Boterenbrood and Nieuwkoop,* 1973]. On the other hand, when the archenteron roof is divided into pieces at the open neural plate stage, only those of the anterior half (which adjoins the yolk mass) produce their normal achievements when cultured in saline: the future prechordal plate gives rise to mesenchyme and the rest forms notochord. The pieces of the other half produce muscle as well, and the most posterior ones even kidney [*Sala,* 1955]. Apparently at the stage of operation the inductive effect has not yet reached its full extent: not all cells have yet been elevated to the level of notochord.

All these data taken together suggest that *the determination and organisation of the mesoderm use only one source of information, the yolk endoderm* (fig. 11). Under its influence the cells are successively determined as lateral plate mesoderm, intermediate mesoderm, somites, notochord, and then probably as prechordal plate and finally as pharyngeal endoderm. At the time competence disappears the cells have 'accumulated' a larger number of determinations as they came under the influence of the inducer earlier. *The result is a spatial sequence of six different tissues, which shows that an environmental prepattern was provided by the yolk endoderm.* There is no antecedent prepattern in this case. *The bilateral symmetry that appears in the mesoderm is established on the basis of an imprinting prepattern: the dorsoventral organisation of the yolk endoderm.* The mid-dorsal region of the embryo obviously is in advance of the rest: it is there that the tissues appear that 'accumulate' the largest number of determinations, and that the blastoporal groove appears; it is there that the marginal zone of prospective mesoderm extends furthest towards the animal pole. All this suggests that there exists a gradient in the yolk endoderm that desynchronises the activities of its cells. The cells seem to acquire their inductive capacity the earlier as they are situated closer to the mid-dorsal region. Instead of being arranged parallel to the endodermal-mesodermal boundary the six prospective areas are in a different pattern, probably one of concentric zones around the prospective pharyngeal endoderm. This would not be apparent on the fate map however, because at the stage to which it refers the induction is not yet completed; it continues while the morphogenetic movements of gastrulation change the relative positions of the cells (see fig. 15).

The Regionalisation of the Neural Plate in Amphibians

When the neural plate becomes visible the prospective areas of prosencephalon, rhombencephalon, spinal cord and tail mesoderm are already determined in it. When they are cut out and grafted to another embryo they form well-identifiable structures: those they would have formed in normal development [*Mangold,* 1933]. This antero-posterior organisation does not reflect some invisible organisation that the ectoderm would have acquired prior to neural induction but is imprinted on it by the archenteron roof. This is clearly shown by the results of transplantations of the dorsal blastoporal lip (p. 31). When this is taken from an early gastrula it consists of cells that will end up in the anterior part of the archenteron roof: accordingly it induces cephalic structures. When it is taken from an advanced gastrula it consists of prospective posterior archenteron roof cells: accordingly it usually induces tail structures [*Spemann,* 1931]. Also, the part of the chordomesoderm that lies below the prospective rhombencephalon induces rhombencephalon in competent ectoderm [*Raven and Kloos,* 1945]. Nevertheless, the organisation of the archenteron roof does not imprint itself on the ectoderm concurrently with neuralisation [*Nieuwkoop* et al., 1952]. In fact, after the future mesoderm has passed around the blastoporal groove it glides below the ectoderm from back to front. Because neural determination requires only a few minutes of contact, neuralisation of the ectoderm is only effected by the most anterior cells of the archenteron roof. As a result of this 'activation' process the ectoderm organises itself into prosencephalon. The regionalisation of the neural plate requires a second induction or 'transformation', which can be experimentally dissociated from the first induction. When the posterior part of the neurectoderm is separated from the archenteron roof immediately after its determination and grafted to the ventral side of the same embryo, it forms prosencephalon, not spinal cord [*Eyal-Giladi,* 1954]. Similarly, ectoderm that self-neuralises in vitro in the absence of an inducer always forms prosencephalon [*Barth,* 1941]. Conversely, the anterior portion of the neural plate develops into prosencephalon when isolated but into spinal cord and rhombencephalon when combined with posterior mesoderm [*Toivonen and Saxén,* 1966; *Toivonen,* 1967].

The differences in transforming capacity along the archenteron roof are not qualitative but quantitative in nature. When competent ectoderm is combined with chordomesoderm from the middle region of the embryo all the regional structures of the neural axis are formed [*Nieuwkoop* et al., 1952; *Sala,* 1955]. Spinal cord is formed in contact with the inducer, prosencephalon next to the part of the ectoderm that escapes neuralisation, and rhombencephalon in between the two. The more or less normal sequence of the various parts of the neural axis in these experiments suggests that the transforming influence decreases as it extends through the neuralised tissue. It also suggests that the transforming capacity of the archenteron roof gradually declines from back to front, which reflects the existence in it of a covert graded pattern. The latter in its turn could be the result of the propagation in the mesoderm of the inductive effect of the yolk endoderm. In fact the prospective prechordal plate can induce spinal cord before it has invaginated [*Okada and Takaya,* 1942; *Hara,* 1961]. The cells would therefore acquire their transforming capacity as they are mesodermalised, and then lose it again as they are 'elevated' to the level of prechordal plate and pharyngeal endoderm. The involvement of this graded pattern in the organisation of the neural axis explains the presence of graded features in it: the morphology of the rhombencephalon is transitional between that of the prosencephalon and that of the spinal cord. However, the graded pattern is not imprinted on the neural ectoderm level by level. In fact one can obtain a complete and well-proportioned neural axis from competent ectoderm by two associated heterogenous inducers: liver, which induces archencephalon and sometimes rhombencephalon, and bone marrow which occasionally induces spinal cord but predominantly mesoderm [*Toivonen and Saxén,* 1955]. Obviously interactions between the various parts of the neural plate play a large part in the establishment of the normal pattern: regionalisation implies a readjustment of the neuralised cells of the neural plate as a whole in response to the presence of the archenteron roof.

The prosencephalon itself constitutes a complex organ whose various parts (telencephalon, diencephalon and optic vesicles) are not yet determined at the early neurula stage. Removal of a part and re-implantation in a different orientation does not lead to abnormal development [*Spemann,* 1918]. Systematic experiments carried out at that stage [*Boterenbrood,* 1970] suggest that the entire prosencephalon possesses a distinct tendency to form eye material. Apparently the formation of the telencephalon is the result of a conversion of the peripheral cells by the non-neuralised ectoderm [*Boterenbrood,* 1958, 1970]. The prospective prosencephalon was cut out together with the adjacent prospective epidermis at the open neural-plate stage, disaggregated and cultured in saline. After reaggregation the neuralised material formed smaller or larger masses surrounded by epidermis. The smallest of these developed entirely into telencephalon. Only the larger ones formed eye material, which was always localised centrally, as if the influence of the epidermal cells had failed to penetrate to the centre.

After extirpation of the archenteron roof an excess of eye material forms in the prosencephalon, resulting in a large single median eye [*Adelmann,* 1937]. The normal division of the eye material into two symmetrical eye anlagen and the harmonious development of the various parts of the prosencephalon require a weak transforming influence of the archenteron roof [*Boterenbrood and Nieuwkoop,* 1961]. In each of the optic vesicles only the outer surface, which comes to lie against the epidermis, will produce neural retina by autonomous progression of differentiation. The remainder develops into pigmented retina because its cells extend on the mesenchyme surrounding the eye [*Lopashov and Khoperskaya,* 1967].

These experimental results, together with the general principles of induction already presented, allow us to view the genesis of the neural axis as a readjustment, the programme of which comprises three separate prepatterns (fig. 12). (1) Neuralisation initiates an autonomous progression which in the absence of other information would lead to the production of eye material. At the same time neuralisation confers on the cells the capacity of spontaneously arranging themselves around a central cavity: *the tubular shape of the neural axis results from an elementary social prepattern.* None of

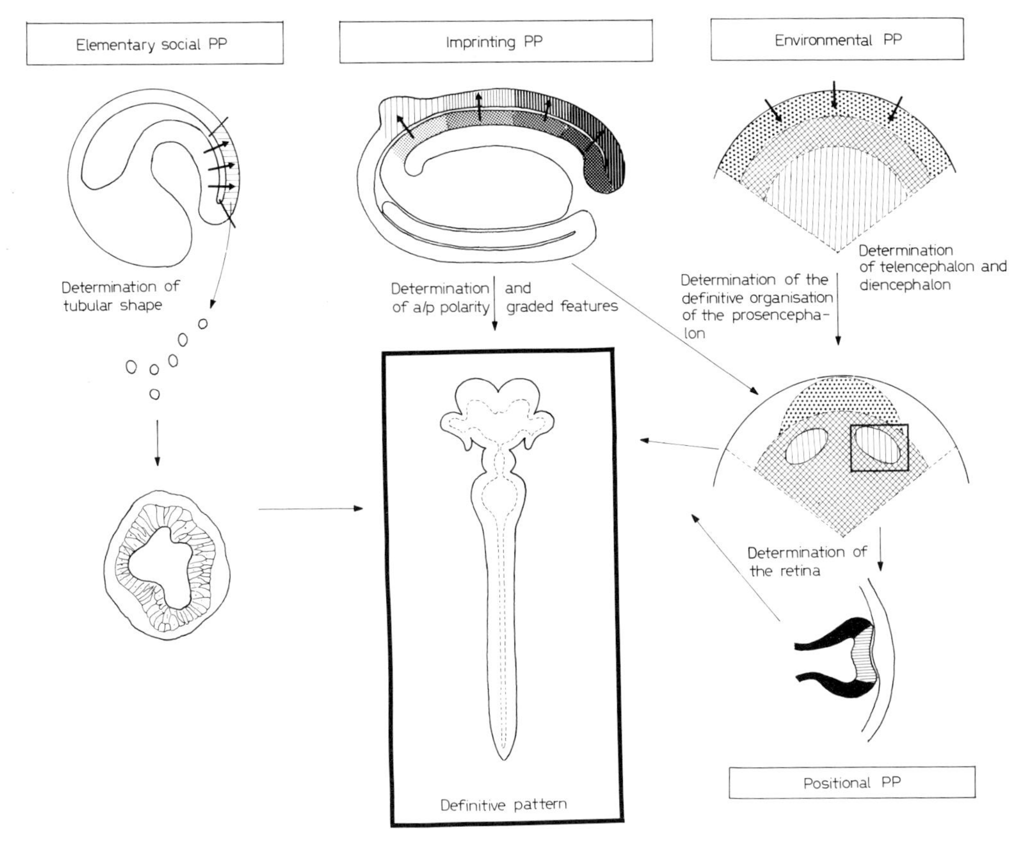

Fig. 12. Interference of prepattern effects in the organisation of the neural plate. Explanation in text.

the features of the further organisation of the tube seems to result from some invisible organisation of the ectoderm existing prior to neuralisation: there is no antecedent prepattern. (2) *The covert graded pattern of induction present in the archenteron roof* (due to the environmental prepattern furnished by the yolk endoderm) *serves as imprinting prepattern.* It deflects the greater part of the neural axis from its autonomous progression and leads to that part losing its capacity of forming prosencephalon. It also manifests itself in the anteroposterior polarity of the neural axis and in graded features of its organisation, in particular the transitional structure represented by the rhombencephalon. (3) *The non-neuralised ectoderm surrounding the neural plate furnishes an environmental prepattern.* In the non-transformed region it deflects the differentiation of the cells which would form eye material in the direction of telencephalon and, more centrally, diencephalon. The features determined by this environmental prepattern and some of those determined by the imprinting prepattern (restriction of the amount of eye material and its division into two symmetrical rudiments) interfere with each other. *The optic vesicles in their turn organise themselves thanks to a positional prepattern furnished by the epidermis and the mesenchyme.*

The Limbs of Amphibians and Birds

Limb formation begins before the emergence of the limb bud by a local condensation of the flank mesoderm that is already determined to form a limb. The overlying ectoderm thickens into the so-called apical ectodermal ridge. The ridge disappears only when the limb is completely formed. It is maintained by a heterotypic action exerted throughout limb formation by the limb mesoderm. If the latter is replaced by mesoderm from another region the ridge regresses [*Zwilling,* 1961]. The action of the limb mesoderm is attributed to the production of an 'apical ectoderm maintenance factor' [*Zwilling and Hansborough,* 1956]. The maintenance of the ridge in its turn is indispensable to limb development: if it is excised before the limb bud becomes visible this never appears [*Saunders,* 1948] (unless the wound epidermis forms a new ridge, which is the case in amphibians but not in amniotes).

This interaction between the two components of the limb bud is sufficient for a normal limb to develop, without further information coming from neighbouring regions. In fact, a prospective limb bud isolated shortly before its emergence and transplanted to the intra-embryonic coelom organises itself in the same manner as if it had been left in place [*Hamburger,* 1938]. Nevertheless, the pattern that appears is a rather complicated one, even if one only considers the formation of the skeleton, as we shall do. During the growth of the bud the mesodermal cells aggregate locally to form precartilaginous condensations, each of which will give rise to a skeletal element. The first condensation to appear is that of the most proximal skeletal element, the stylopod (humerus in the forelimb); it appears shortly after the emergence of the bud. Meanwhile the bud elongates due to the activity of an apical zone of proliferation where mitoses are more numerous. This zone produces mesoderm that comes to lie between the stylopod and the apical ridge. In it appear two further precartilage condensations, the rudiments of the two elements of the next limb segment, the zeugopod (radius and ulna in the forelimb). The limb continues to elongate apically by the same process but now it broadens and flattens dorso-ventrally, forming a hand (or foot) plate representing the future autopod. Several rows of precartilage condensations appear, starting proximally, which represent the five digital rays. The rays do not form simultaneously but one after the other, the most posterior one (digit V) appearing first.

We have already seen that the formation of precartilage condensations is related to the elementary social behaviour of cells that are determined to give rise to limb mesoderm. In vitro the number of condensations is proportional to the volume of the explant [*Ede and Flint,* 1972]. This phenomenon is the key to understanding the arrangement of skeletal elements in the proximo-distal direction [*Ede and Agerbak,* 1968; *Ede,* 1971; *Ede* et al., 1977]. When the limb bud first appears it contains but little mesoderm and consequently only a single precartilage condensation is formed. When the bud has grown to a certain length condensation of the newly formed mesoderm is again possible, but because the distal end is now beginning to broaden two condensations will appear. In the hand plate the number of condensations will again be higher. In this way the proximo-distal organisation of the limb is related to apical growth. The latter is maintained by the apical ectodermal ridge. When this is removed from the chick limb bud (where it usually does not regenerate) only those limb segments will later be present and normally differentiated which were already determined at the time of the operation [*Saunders,* 1948]. Conversely, if one grafts an extra ectodermal ridge to the base of an early chick limb bud the limb will later be bifurcated distal to the level of implantation [*Zwilling,* 1956]. The growth of explants of limb mesoderm is stimulated by association with apical ectoderm [*Gumpel-Pinot,* 1980], even when the latter is first disaggregated [*Saunders,* 1977]. Incidentally, mitotic activity in the apical zone of proliferation seems to be insufficient to account entirely for the formation of the hand plate. In addition the apical ridge would induce the mesodermal cells to migrate distally to accumulate in the apical region of the bud [*Ede and Agerbak,* 1968]. It is interesting to mention that this hypothesis was confirmed by a computer simulation of limb growth [*Ede and Law,* 1969].

When it has reached the hand plate stage the limb has an antero-posterior polarity reflected in the asynchronous appearance of the digits, and a dorso-ventral polarity expressed by its distal part being flattened in the frontal plane. These axes of polarity reveal an organisation that already existed before the limb bud appeared but is bound up with invisible features of the cells' individualities. To answer the question whether it resides in the mesoderm or in the ectoderm, chick limb bud mesoderm was isolated and 'minced' and then recombined with an intact ectodermal hull. The whole was grafted to a host embryo. The differentiated limb showed distinct dorsal and ventral surfaces but its anterior and posterior edges were similar and atypical [*MacCabe* et al., 1973]. Therefore, the two polarity axes obviously correspond to two different prepatterns. The d/v polarity resides in the ectoderm and is transferred from there to the mesoderm. The a/p polarity, on the other hand, resides in the mesoderm, but because this cannot develop by itself it transfers its polarity to the ectoderm, which then 'acts back' on the mesoderm. In fact the posterior part of the ectodermal ridge is the first to thicken, and is probably responsible for the early appearance of the precartilage condensation of digit V. This two-way interaction has been confirmed by the following experiment. The apical part of an early wing bud is cut off and placed back on the stump after 180° rotation. The thin, originally anterior part of the apical ridge thickens, obviously under the influence of the posterior mesoderm of the stump. Later a posterior digit appears there, followed by more anterior digits. The originally posterior part of the ridge remains thick under the influence of the subjacent posterior mesoderm. A posterior digit appears there also, followed by other digits, so that ultimately the wing will be reduplicated, the anterior hand showing a reversed antero-posterior polarity. One has concluded that there exists in the mesoderm an antero-posterior gradient [*MacCabe* et al., 1977] which manifests itself in the distribution of the maintenance factor [*Saunders* et al., 1958; *Saunders and Gasseling,* 1968].

The two axes of the limb, which correspond to two separate prepatterns, are not determined at the same stage. While in the amphibians one can still invert the d/v axis by rotating the prospective limb area through 180° just before the emergence of the bud [*Harrison,* 1921], the a/p axis cannot be reversed in this way even at the gastrula stage [*Detwiler,* 1933].

In conclusion, then, the unfolding of a spatio-temporal pattern in the mesoderm of the limb is due to four different prepatterns (fig. 13): (1) *The formation of the many cartilage elements which will constitute the limb skeleton is due to an elementary social prepattern* established when the mesoderm is first determined as limb mesoderm. (2) *The specific arrangement of skeletal elements in a proximo-distal direction is due to an environmental prepattern originating from the apical ectodermal ridge,* which in its immediate neighbourhood modifies the elementary social behaviour of the mesodermal cells (cell division and cell migration). (3) *The antero-posterior polarity reflects an invisible graded pattern that is present as early as the gastrula stage and serves as an antecedent prepattern.* (4) *The dorso-ventral polarity is determined by an imprinting prepattern originating from the ectoderm.*

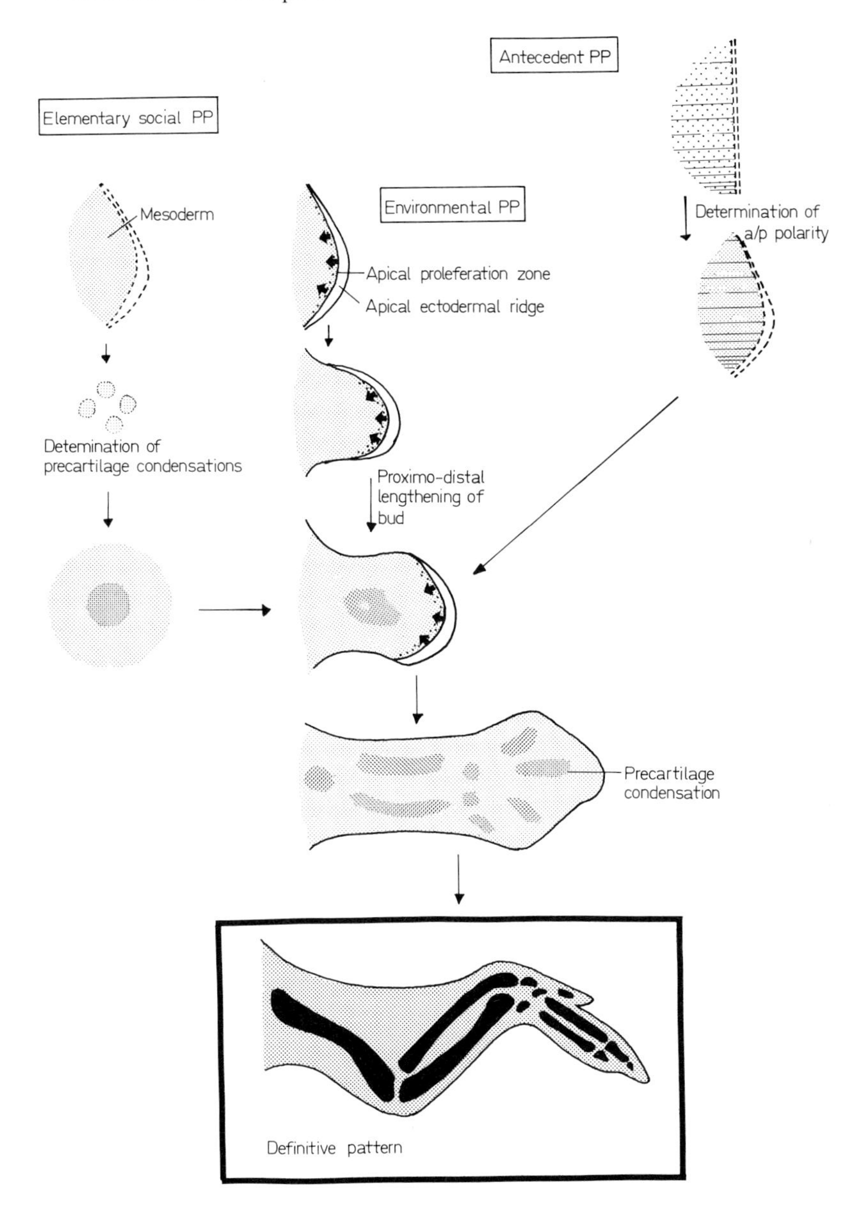

Fig. 13. Interference of prepattern effects in the organisation of the limb. Explanation in text (the imprinting prepattern which specifies the d/v polarity of the bud is not shown).

Summation of Pattern Features

The pattern established in a cell population during a readjustment serves as antecedent prepattern for the next readjustment and as imprinting prepattern for the readjustments of neighbouring cell populations. From the beginning of functional organogenesis onwards further readjustments are triggered by morphogenetic hormones. The postembryonic spatio-temporal patterns of growth are specific readjustments resulting from changes in the humoral constellation. These readjustments only concern mitotic activity and use as antecedent prepatterns the covert graded patterns established during pre-functional organogenesis. The period during which a prepattern is established coincides with a 'critical period' during which non-specific factors, which alter the elementary social behaviour of the cells, can determine definitive aberrations in organisation.

The structuring of the organism during development reflects an enrichment of the collective memory of its cell populations. This is a social phenomenon that has its origin in the capacity of the cells to enrich their individual memories. Because each cell at any time interprets the extracellular information that it receives in terms of the information it had stored previously, the cell population as a whole interprets each prepattern in a way which depends on the contents of the successive prepatterns which previously had been provided to it. In other words, because cellular differentiation proceeds by the summation of extracellular information, the reproduction of structural patterns proceeds by the summation of the effects of prepatterns. In particular, each time a cell population engages anew in a readjustment as a result of a determinative event, the pattern that had appeared during earlier stages serves as an antecedent prepattern, and at the same time as imprinting prepattern for neighbouring cell populations (fig. 14).

An instructive example is furnished by the organisation of the somites in birds, with its repercussions for feather formation from the epidermis. Somite formation results from a change in the elementary social behaviour of the mesodermal cells: their adhesiveness increases [*Bellairs* et al., 1978] and the properties of the intercellular material change [*Bellairs,* 1979]. To form a somite a certain number of cells autonomously arrange themselves

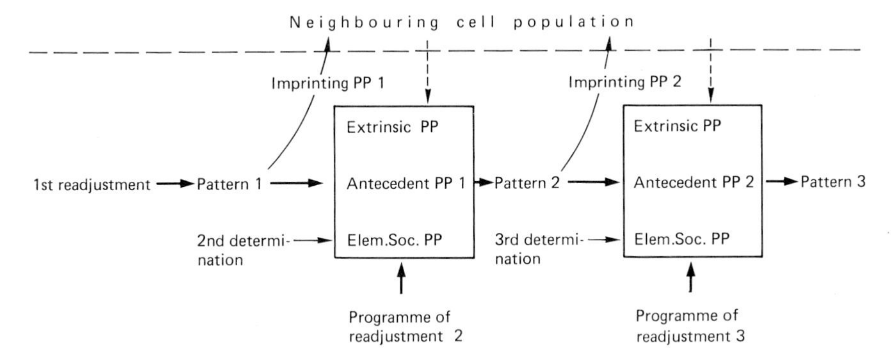

Fig. 14. Diagram showing the summation of prepattern effects. Explanation in text.

around a small central cavity. The laying-down of rows of somites would be impossible when all the cells would be determined simultaneously: it proceeds from front to back, a somite forming each time a sufficient number of cells change their social behaviour. Morphological differentiation is preceded by two successive waves of determination which also propagate in antero-posterior direction and play different roles in the establishment of somite patterns [*Elsdale and Pearson,* 1979]. The first wave occurs at the blastula stage and corresponds to the propagation of the induction emanating from the yolk endoderm (p. 73–74). It conditions the desynchronisation of cellular activities that occurs at that stage. The further the cells are away from the inducer, the later they become determined and will start rearranging themselves. The second wave occurs after gastrulation and corresponds to an 'elevation' to the level of somitic mesoderm under the influence of the notochord (p. 73). Because the most posterior cells are the last to arrive next to the notochord they are the last to be determined in this way. The cells of the early somites are engaged in an autonomous progression leading to the production of muscle tissue. However, this progression is deviated in the neighbourhood of the notochord, where the somite produces cartilage [*Holtzer,* 1952], and in the vicinity of the epidermis, where it produces dermis [*Mauger,* 1972].

We may summarise these results (fig. 15) by saying that at the blastula stage a graded pattern of induction is established under the influence of the yolk endoderm. It serves as antecedent prepattern for the layout of the somites. At the neurula stage this antecedent prepattern interacts with the elementary social prepattern which was provided to the mesodermal cells when they were 'elevated' to the somitic level by the notochord. From this results the segregation of the somitic mesoderm into somites. Each somite

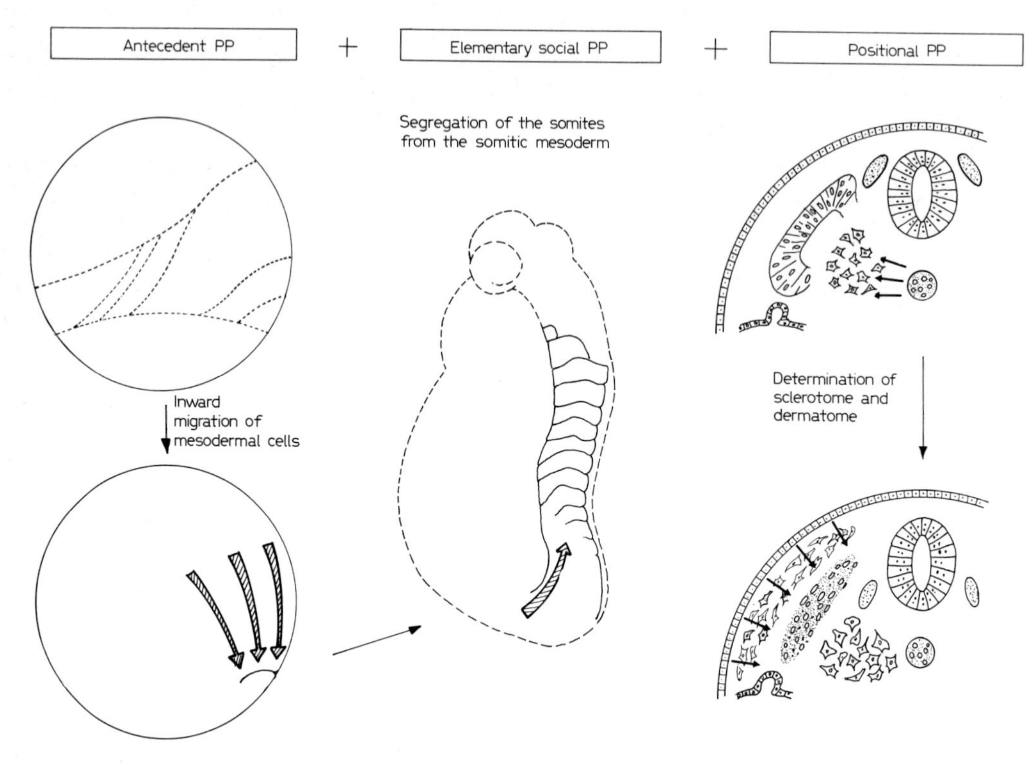

Fig. 15. Summation of prepattern effects in the differentiation of the somites. Explanation in text.

then responds to a positional prepattern originating from the notochord and the epidermis.

One of the consequences of these transformations of the somitic mesoderm is the appearance of the plumage of the dorsal body surface (fig. 16). The embryonic feathers are purely ectodermal structures. Each of them is induced by a cell condensation in the underlying mesenchyme. The feather germs form in a certain number of distinct tracts, within which they are arranged in regularly alternating rows, forming a 'hexagonal' pattern. This pattern is imprinted on the ectoderm during feather induction by the pattern of cell condensations in the dermal mesenchyme (imprinting prepattern) [*Sengel and Rusaouën,* 1968]. Its establishment has been studied for the so-called spinal tract, which occupies the major part of the dorsal surface of the embryo. The underlying mesenchyme originates from the dermatomes, i.e. the portions of the somites that are in contact with the epidermis [*Mauger,*

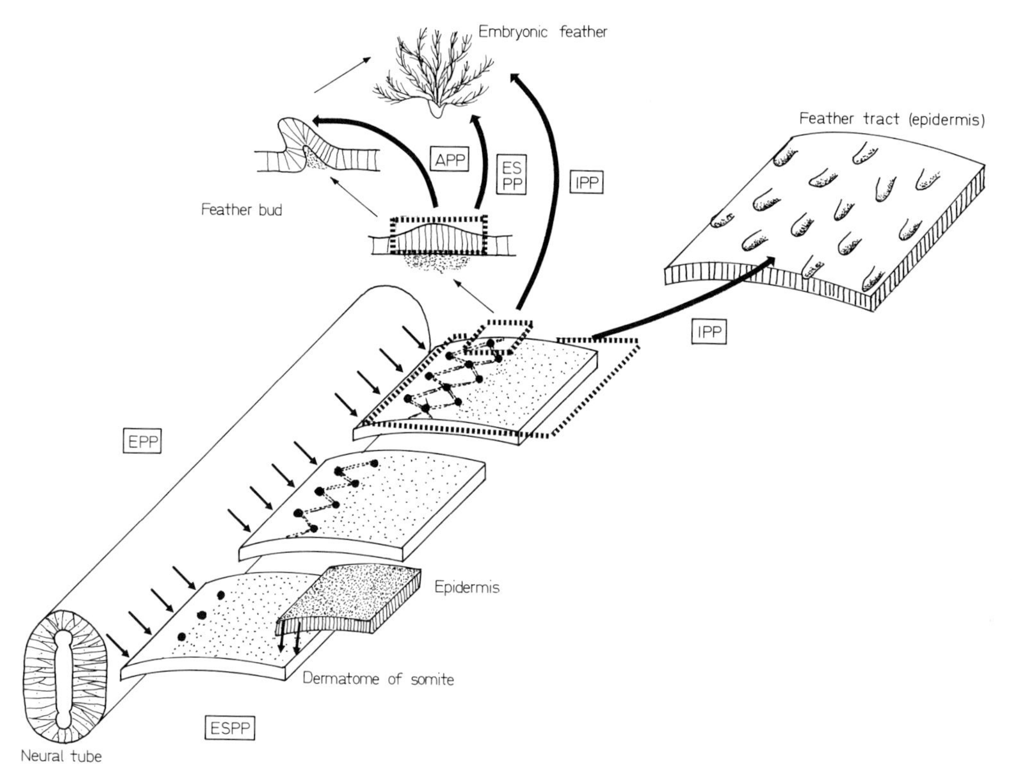

Fig. 16. Interference and summation of prepattern effects in the determination of the embryonic feathers and feather tracts. Explanation in text. APP = Antecedent prepattern; EPP = environmental prepattern; ESPP = elementary social prepattern; IPP = imprinting prepattern.

1972]. It is probably as a result of the determinative event resulting from contact with the epidermis that the mesenchymal cells acquire a new type of elementary social behaviour. Theoretically it seems impossible that the resulting cell condensations could appear all at the same time [*Ede,* 1972], and in fact they do appear row after row, the appearance of one row evoking that of the next one. Only for the first (most median) row is a sort of induction by the neural tube required; its anterior condensations are the first to appear [*Sengel and Novel,* 1970]. In other words, the appearance of the condensations is due to the interference of two prepatterns, an elementary social and an environmental one. The overall shape of the feather tract is thought to be connected with the antero-posterior polarity of the somitic mesoderm [*Mauger,* 1972].

The embryonic feathers themselves have a complex organisation which varies among body areas and species. At their tips they carry branches called barbs, whose number, shape and size are determined by the nature of the mesenchyme – for instance, recombinates of femoral mesenchyme with wing epidermis form feathers of femoral type. Therefore, the mesenchymal condensation provides an imprinting prepattern to the feather it induces. The other features of the embryonic feather are determined by intrinsic prepatterns. The backward slant of the germ is due to the antero-posterior polarity of the ectoderm, which is fixed as early as 24 h of incubation [*Sengel and Mauger,* 1976] and is used as an antecedent prepattern. The number and arrangement of the barbules, the side branches of the barbs, are determined by the nature of the epidermis (elementary social prepattern).

In the course of their differentiation the cell populations acquire new properties. Consequently, the successive readjustments they effect are not of the same nature throughout development (even though they are based on the same principles as the embryonic readjustments). Organogenesis begins by homotypic interactions and heterotypic actions occurring exclusively among topographically adjacent cells. This mode of coordination of cellular activities we have earlier called the 'primary integration system' [*Chandebois,* 1976a, b]. However, as we have seen on page 49 the exchange of positional information between cells is not sufficient for the cells to complete their definitive differentiation and to continue their growth. Without the influence of morphogenetic hormones a premature state of equilibrium would be attained in the whole organism. The hormones provide to the cells new extracellular information which is added to or modifies the previous positional information. Thus, a new mode of coordination of cellular activities takes over: the 'secondary integration system'. As a first result the specific biosynthetic activities of the cells are modified: the tissues either complete their differentiation, or 'modulate', or their cells cytolyse. All these processes constitute what we call 'functional organogenesis'. But a time will arrive when changes in the luxury metabolism of the cells are no longer possible. If we disregard the appearance of the secondary sex characteristics that follow the maturation of the gonads, new features no longer appear in the structural patterns. From this time on only mitotic activity continues to be controlled by the morphogenetic hormones: this is the period of 'postembryonic growth'.

In all cells the cytoplasmic information content determines how the cell will react to a hormone. Therefore, the nature and sequence of the transformations resulting from the activity of the secondary integration system depend to a very large extent on the organisation that was established due to the activity of the primary integration system. In particular, covert graded patterns will now manifest themselves, sometimes for the first time, in a desynchronisation of cellular responses. For instance, cytolysis in the metamorphosing tail, both in vivo and in vitro [*Weber,* 1963], follows a spatio-temporal pattern that reveals various pre-existing graded patterns. The tail fins are progressively resorbed (first the dorsal, then the ventral one) beginning with the free border, followed by cytolysis of the body of the tail, which begins at the tip and proceeds more rapidly dorsally than ventrally. Similarly, during postembryonic growth changes in the proportions of the organism occur which sometimes amount to veritable distortions. If one maps out the growth of different parts of an organ with the aid of an arbitrary set of coordinates (for instance a rectangular grid) one always finds a centre of intense growth from which the growth rate regularly declines in all directions. In other words, such methods demonstrate the existence of 'gradients of allometric growth' [*Huxley,* 1932]. Without doubt the spatio-temporal patterns characteristics of postembryonic growth are the manifestations of covert graded patterns established during prefunctional organogenesis, when the metabolic standing and hence the mitotic rhythm of the individual cells become infinitely diversified. This phenomenon is not a new one in development, for cell division becomes asynchronous as early as the beginning of cleavage, but it is only during this period that it is no longer masked by the diversification of cell individualities and hence becomes well accessible to mathematical analysis [*von Bertalanffy,* 1960].

A cell is only capable of responding in a certain manner to a morphogenetic hormone if its concentration exceeds a threshold, the level of which is rigorously fixed by the cell's cytoplasmic information. Consequently, *this whole second period of development (when the secondary integration system is functioning) consists of successive phases, each of which is determined by a particular 'humoral constellation'.* During amphibian metamorphosis, as the activity of the thyroid increases, the various organs become transformed in the order of increasing threshold levels of response to thyroxine [*Etkin,* 1935]. Postembryonic growth also shows successive phases, each of them characterised by a particular growth rate of the whole organism and of the various organs [*von Bertalanffy,* 1960]. The transition of one phase to the next is usually irreversible and abrupt, and connected with a change in physiological state [*Abeloos,* 1949].

When the cells of a cell population (or some of them) transform their individualities as a result of a change in the humoral constellation, they do not do so independently of one another. The action of the hormone, which is equivalent to a determinative event, confers on them a new elementary social behaviour, which leads them on to a new readjustment. Hence the pre-existing features of organisation are necessarily also transformed. Later, as the humoral constellation changes further, a chain of readjustments results on the same principles as during prefunctional organogenesis. In other words, *the secondary integration system does not take the place of the primary system but 'rewinds' it when it has 'run out', thus allowing the completion of development.* As long as the specific activities of the cells can still be changed by the action of the morphogenetic hormones, the resulting readjustments necessarily lead to further structural complication and organogenesis proceeds. The organisation established at the end of a readjustment serves as antecedent prepattern when the cell population itself undergoes a new determinative event, and as imprinting prepattern when an adjacent cell population becomes determined anew. The summation of prepattern effects continues as in prefunctional organogenesis. This necessary concatenation of events is demonstrated by experiments performed on young frog tadpoles. It is impossible to obtain simultaneous metamorphosis of all organs by a single injection of a high dose of thyroxine; the resulting adults will be highly abnormal. The only way to ensure normal metamorphosis is a series of injections of increasing dose [*Etkin,* 1935]. Apparently the response of certain cell populations to low concentrations of thyroxine prepares the way for a 'normal' response of those cell populations that have a higher threshold level.

Once the histotypic activities of the cells are terminally specified and stabilised, changes in the humoral constellation only act further on mitotic activity. The abruptness and irreversibility of the effects of such changes at the level of cell populations (reflected in changing growth rates) make them equivalent to inductive events. Here again, the cells do not respond individually, for mitotic activity is an aspect of the elementary social behaviour of the cells. There inevitably follows a readjustment, entailing changes in the features of the mitotic pattern: one does indeed find that the profiles of the growth gradients are not comparable from one phase to the next; sometimes they are indeed reversed outright [*Abeloos,* 1949].

At any time in the development of an animal an important aspect of the activity of the two systems of integration (primary and secondary) is to preserve the collective memory of the cell populations of the organism. Thanks to the continuous exchange of positional information, and later the diffusion of morphogenetic hormones through the circulation, the cells permanently retain all the features of their individuality that they had at first acquired in reversible form. Likewise, in cell populations undergoing renewal a continuous registration of information ensures the correct integration of the newly produced cells into pre-existing structures. Consequently, except in the case of trauma or of profound physical disturbances, *the features of organisation that appear after a prepattern has been registered are irreversibly fixed.* If a cell population fails to register a prepattern at the proper time, or if this prepattern is altered, abnormal features of organisation are bound to appear. *By dint of the interplay of successive readjustments these features are bound to become amplified and to reflect on adjoining cell populations.* Nevertheless, even if an organ shows abnormal features due to a change in a prepattern it will usually remain recognisable thanks to the presence of normal features determined by other prepatterns involved in its formation.

These principles can be applied in understanding the teratogenetic effects of various treatments (injection of toxic substances, irradiation, temperature shocks, ultrasound treatment, etc.). These treatments do not necessarily interfere with the specific activities of the cells but more often have commonplace effects such as cell death, changes in cell adhesiveness or changes in mitotic activity. *If a treatment administered to the organism as a whole affects certain cellular activities or properties at the time they are involved in the establishment of a prepattern, the prepattern either will not appear at all or will be altered in some way.* In both cases the result will be a localised and definite structural alteration. To mention an example, after injection of hydrocortisone into the chick embryo the cells of the dermatomes fail to condense. Because there consequently is no imprinting prepattern for the spinal feather tract (p. 84) this will not appear [*Sengel*, 1971]. Another example: administration of aspirin to the pregnant mouse on the 11th day of gestation frequently leads to polydactily. The cell death normally occurring in the apical ectodermal ridge at that time is retarded, as a result of which the ridge persists beyond the normal time. It can therefore act longer on the mesoderm, leading to the appearance of extra digits. It is thought that aspirin enhances the production of maintenance factor by the mesodermal cells (p. 78) [*Klein* et al., 1981].

Because of the commonplace nature of the cell properties or activities that are affected by such treatments, the nature of the treatment is much less important than the stage at which it is applied. For instance, X-irradiation has exactly the same effect on the formation of the chick spinal feather tract as hydrocortisone. To obtain the complete absence of feather germs treatment must be performed on the 5th day of incubation. With later treatment some rows of mesenchymal condensations are already present. They will not be altered and only the more lateral rows of feather germs will fail to form [*Sengel,* 1971]. For each type of anomaly there exists a 'critical period' [*Morris,* 1979] during which the treatment produces an effect.

In connection with the manner in which prepatterns are established and 'interpreted', in the same rudiment several critical periods can sometimes be demonstrated to exist for the same teratogenetic factor, which are separated by periods of insusceptibility. In each of these critical periods the altered elementary social behaviour of the cells interacts with a different specific organisation, so that the deflection of the readjustment each time leads to a different type of anomaly. This has been shown particularly for the formation of the somites in amphibians [*Elsdale and Pearson,* 1979; *Pearson and Elsdale,* 1979]. If embryos are subjected to a temperature shock after the appearance of the first somites, the next few somites will form normally but a certain number of more caudal segments will show anomalies. At this stage the treatment apparently alters the elementary social behaviour of cells already determined to participate in the formation of somites. If, on the other hand, a temperature shock is given at the blastula stage, anomalies appear without apparent order among the 30 most anterior somites. It seems that at that stage the treatment disturbs the gastrulation movements. The cells of the prospective somite area, whose degree of differentiation varies with their distance from the yolk endoderm (p. 73–74), would arrange themselves in disorderly fashion on either side of the notochord. It is remarkable that temperature shocks given at the gastrula stage have no effect whatever.

The possibility of obtaining various different patterns starting from one and the same antecedent prepattern is widely used in normal development as well, namely for the diversification either of structures in one animal or of forms within one species (polymorphism). For instance, in the insects the various body segments with their respective appendages probably derive from analogous antecedent prepatterns. In holometabolic insects the fall in the level of juvenile hormone in the blood of the larva just prior to metamorphosis allows the adult organs to differentiate. Those of different segments take on a different structure. That mitotic activity plays a role here is suggested by the well-known experiments of *Hadorn* [1966] on *Drosophila.* When imaginal discs (the rudiments of the adult organs) are implanted directly into the abdomen of adult flies after being transected, they continue to proliferate without differentiating. They can thus be cultured in vivo for years, by serial transfer of tissue fragments from one host fly to the next. When a piece of an imaginal disc is implanted directly into a mature larva it metamorphoses along with the host and differentiates normally. When this is done with a piece that has first undergone several transfers in the adult abdomen, however, it may differentiate partially or entirely into tissues belonging to another imaginal disc, depending on the duration of in vivo culture. For instance, the genital disc first produces head and antennal tissue, then, after longer proliferation, leg tissue, wing tissue, and finally thorax tissue. This phenomenon is called 'transdetermination' (because we have to do with determined but undifferentiated tissue). Direct investigation of the mitotic activity of transplants has confirmed that transdetermination becomes more frequent as mitotic activity is higher. Other experiments will be described in the next section which corroborate this interpretation of *Hadorn's* [1966] findings.

Numerous animal forms are characterised by sexual dimorphism, and sometimes polymorphism, depending on the environment of the individual or on its integration into the animal society. Nonetheless, all individuals originate from identical embryos. It is only much later, when the organism goes through a period of 'ontogenetic competence' [*Chandebois*, 1976a], that a choice is made between two or more forms which have equal chances of being realised. In all such cases the choice depends either on the humoral constellation, i.e. on the nature and concentrations of certain hormones (produced as a result of stimulation of certain sense organs), or on the presence of 'pheromones' in the blood (pheromones are substances released by other individuals and interiorised through the food or otherwise). Depending on the stage at which competence appears it is either the course of organogenesis or the pattern of postembryonic growth that diverges. Thus, termite larvae give rise to various adult forms corresponding to as many 'castes' of the termite society (soldiers, workers, king and queen). In the male of the crab *Uca* one of the two pincers develops disproportionately. Sometimes the divergence is so great that one would hesitate to place the individuals of one species in the same taxonomic group. This is the case in the echiurid *Bonellia*, where the female is large and has a long proboscis terminating in a double lobe, while the male is tiny and ovoid in shape and develops from a larva that attaches itself to the female; larvae treated with dried female extract develop into males.

It should be pointed out that the period during which the ontogenetic competence manifests itself does not differ from a 'critical period'. In both cases we have to do with a particular phase in development during which the course of a readjustment can deviate in a certain manner under the influence of a factor of external origin.

Gene Expression in the Reproduction of Structural Patterns

The genes involved in the reproduction of structural patterns only play a role in the determination of the elementary social behaviour of the cells – and consequently in the establishment of elementary social prepatterns and environmental prepatterns. A gene mutation – as well as the effects of various (non-mutagenic) treatments – only manifests itself in visible structure from the time its products affect cellular activities bound up with the establishment of prepatterns ('epigenetic crisis'). Because of the interference and subsequent summation of the features imparted to cell populations by prepatterns, the amount of information contained in the DNA can be shown to be sufficient to account for the complexity not only of the structures of the adult but also of the spatio-temporal patterns of development.

The fact that the modes and chronology of the reproduction of structural patterns are so rigorously fixed for every species suggests that the activity of the genes engenders a four-dimensional organisation. The intriguing problem of how this comes about, which is of fundamental import for all areas of biology, has posed itself so early in the history of biology that answers have been formulated before a sufficient body of experimental data was available. The conclusion that will be drawn in this chapter will lead to ideas which deviate considerably from those that today still constitute the basis of classical biology. That is why we will only expound our ideas after having explained first why the problem should be reconsidered from its very fundamentals.

Certain gene mutations lead to more or less severe anomalies in visible structure. If one only considers adult morphology this leads one to think that each particular morphological trait permanently depends on a particular gene. It is because of this misconception – seemingly confirmed by the so-called chromosome maps – that even today one still often subconsciously adheres to Mendel's one gene-one trait relationship, although the biochemists have unambiguously shown that the relation is: one gene-one polypeptide. In fact theorists even today hold that the three-dimensional organisation of an organism at a given time represents the *instantaneous* transposition of the linear sequence of the nucleotides in the DNA. But what could be the principles underlying such transposition? Moreover, every mutation is expressed at a specific stage earlier or later in development. This fact, interpreted on the basis of the Mendelian gene concept, has greatly contributed to the credence given to the theory of sequential gene derepression. But if this were true, what hierarchy in the DNA sequences (presumably established during evolution) and what complicated strategy laid down in the genetic 'programme' could make the genes coding for the finer structural detail express themselves last? Among the numerous problems for which an erroneous conception of morphogenesis has sought purely speculative solutions is the one that interests us here: the automatism of development. In the current view the organism throughout its existence is seen as a machine that decodes the genetic information contained in the DNA and is comparable to the machines used in telecommunication for the transmission of messages. The amount of information in the 'decoded' message is the same as that in the 'encoded' message. However, it soon became clear that the number of genes in animals is far insufficient to account for the complexity of the adult [*Ashby*, 1958] – not to mention the complexities of organogenesis. The problem becomes even more disturbing if one realises that most of the genes are redundant and that there is no relation between amount of DNA and position on the evolutionary scale.

Some authors, for whom there was no need to call the usual gene concept into question, have proposed to look to the environment for 'complementary organisation factors'. This idea is at the basis of a theory formulated by *Atlan* [1972]: the principle of development would be analogous to that of the 'emergence of order from noise'. Although the environment does not furnish any specific information ('the animal is closed to information' [*Apter*, 1966]), it would contribute to the amplification of the structural complexity of the organism. This theory lacks any experimental foundation. Other authors [*Apter and Wolpert*, 1965; *Apter*, 1966; *Arbib*, 1972] already proposed to replace the idea of a 'decoding' of the genetic information by that of 'instructions' comparable to the input data supplied to automata or to a computer. However, here again the true problem remained unsolved because the authors always consider the DNA to contain the programme for the activities of the cell.

Such concepts appear less and less capable of being maintained in the light of recent progress in molecular biology, for the inflexible concept of gene function they imply is collapsing (p. 13). Initiation and termination sites are probably not rigidly fixed. The template for the translation of a polypeptide is probably a mosaic of mRNAs transcribed from different sites or even chromosomes. Somatic cells are probably incapable of reactivating completely repressed genes. The metabolic specialisation of cells would in essence be based on quantitative differences rather than on the repression of parts of the genome that would no longer be needed for the diversification of cell individualities. In all these cases it has been established that physico-chemical conditions in the extrachromosomal milieu are determinative for gene function and for the utilisation of gene products. All these new data have prompted *Scherrer* [1980] to proclaim that 'the genes are non-existent as physical entities in the DNA' and to propose a model, called 'cascade regulation', to explain the 'dispersion of the gene by multistep posttranscriptional controls'. It is clear that it is high time to propose new concepts if we do not wish to see embryology and molecular biology going entirely separate ways, which must be dead ends because they are separate.

The principles of development which emerge from the concept of cell sociology can be argued to be in substantial agreement with this revolution in our knowledge of gene function. To clarify the discussion, let us imagine a segment of DNA (gene A) involved in mRNA synthesis (fig. 17). The sequence in question will of necessity be transcribed from the very start of development but the complementary RNA sequence will not always be incorporated into the same type of messenger. As the cytoplasmic information content changes during the progression of differentiation, the initiation and termination sites could be displaced and various different mRNA mosaics could be formed, which would lead to changes in the composition of protein families. Moreover, as new molecules appear in the extrachromosomal milieu (produced by other genes of the same cell or by other cells) the polypeptides will be inserted into complexes (structural molecules or enzymes) of novel types. This latter process has been extensively discussed by *Hadorn* [1958], who calls it 'stepwise insertion of gene products'. If we add to this that the presence of a particular molecular complex only becomes manifest in patent structure beyond a certain concentration threshold, we can say with certainty that a visible modification of the individuality of a cell does not reflect the concomitant activation of a transcription site.

During development a cell cannot acquire new individual features nor retain those it has already acquired if it is isolated from other cells. On the other hand, not all features of a cell's individuality enter into the determination of visible structure. Consequently, the appearance of new molecular

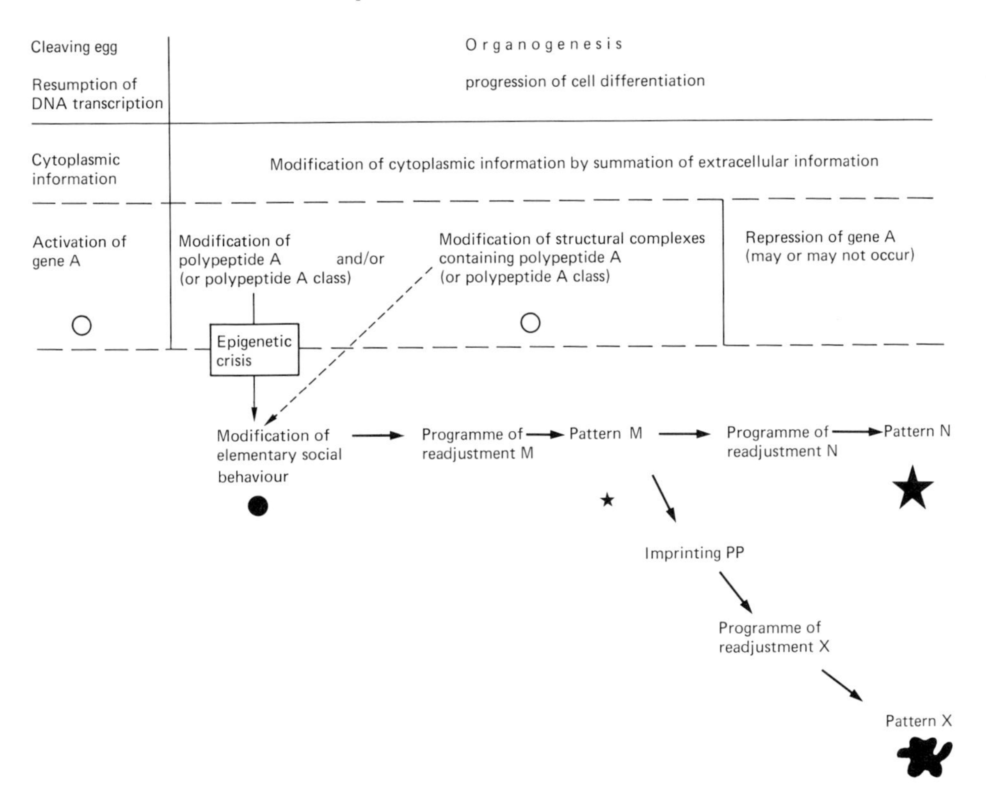

Fig. 17. Diagram summarising the effect of a gene A on the reproduction of structural patterns N and X (see text). The effects of a mutation (open circles) of this gene do not become visible before the 'epigenetic crisis'. After the crisis visible alterations of pattern features appear (solid circles) even when gene A is repressed. These alterations will be amplified from one readjustment to the next (other symbols).

species in cells can only have a morphological effect if they modify an aspect of the elementary social behaviour of the cells, and if that behaviour is involved in the establishment of a prepattern. In support of this idea we may cite the work of *Wilde and Crawford* [1966], who found that a treatment that interferes with protein synthesis in early cleavage stages of the fish embryo does not have immediate visible effects; the effects are delayed until after gastrulation and then manifest themselves in abnormal morphogenesis. The notion of 'stepwise insertion of gene products' is insufficient because it only considers events at the level of the single cell. In reality *it is through the interference and summation of features determined by prepatterns that the genes control the modes and chronology of the reproduction of structural patterns during development.*

Let us go back now to the segment of DNA we introduced earlier. Let us assume that as a result of a mutation it contains an atypical nucleotide sequence (fig. 17). The whole class of polypeptides produced under its control will be atypical and will cause anomalies in the structural complexes into which they are incorporated (or, in the case of enzymes, in the substances whose production they catalyse) – and all this right from the start of development. But the mutation will not express itself in visible structure before these atypical molecules become involved in the establishment of a prepattern. In those cases where the affected cell population in question can still engage in a new readjustment, indelible abnormal pattern features will appear; subsequently they cannot but be amplified from one readjustment to the next, even if transcription on the mutated DNA segment has ceased long since. As an example of such a chain of effects we will once more consider the mutation *talpid*^{3} of the chick. It enhances the adhesiveness of mesodermal cells and consequently reduces their motility. When the limb bud organises itself the number of cartilage condensations formed per unit volume is higher than normal and thus the elementary social prepattern is altered. Moreover, the cells move more slowly towards the apical ectodermal ridge. The environmental prepattern contributed by the ridge is not correctly interpreted and the foot or hand plate is not properly individualised. In the fully differentiated limb the stylopod and zeugopod are shortened and there are too many digits. All the mesodermal derivatives in the embryo are affected in the same manner by this mutant. The formation of the dermal condensations which establishes the imprinting prepattern for the feather tracts is disturbed: hence anomalies are observed in the feather pattern [*Ede,* 1972]. Distortions also occur in the facial area [*Ede and Kelly,* 1964].

Because the effect of a mutation is often localised in a restricted area it has sometimes been thought that the function of the genes is to control the structure of a more or less strictly delimited area. However, an old observation in *Drosophila* [*Waddington,* 1939] already shows that the phenomenon is linked up with the localisation of prepatterns. Flies carrying the 'homoeotic' mutation *aristapedia* have antennae that partly show the organisation of a leg. Animals carrying the mutation *fourjointed* have legs that are reduced in size. If one combines the two mutations in one animal the part of the antenna that has been transformed into leg shows the same reduced size as the legs proper.

Because the reproduction of structural patterns implies the summation of the effects of successively established prepatterns, during the development of an organ at intervals brief periods occur during which morphogenesis can be deviated by the effect of a mutation. *Waddington* [1956] has called them periods of 'epigenetic crisis' and has shown that during the differentiation of the wing rudiment in *Drosophila* there are four such periods, involving six different genes. However, the alteration of a prepattern is bound up with aspects of the elementary social behaviour of the cells, involving ordinary cell properties and activities (such as adhesiveness and cell division). It is therefore not surprising that external (non-mutagenic) factors having similar effects on the cells can be equally decisive during such a period, and can produce so-called 'phenocopies'. It appears there is no reason to distinguish the periods of 'epigenetic crisis' from the 'critical periods' mentioned earlier, because the target of gene action and teratogenetic factors is the same, i.e. the elementary social behaviour of the cell population. For instance, a sublethal temperature shock administered to the wing rudiment of *Drosophila*, shortly before the time wing development can be deviated by the effect of a mutation, produces an effect that is similar to that of the mutation [*Waddington*, 1956]. The work done on *aristapedia* mutants clearly shows that in both cases it is the elementary social prepattern that is altered. Application of various non-mutagenic agents to last-instar larvae of *Drosophila* [references in *Gehring and Nöthiger*, 1973] has resulted in phenocopies of *aristapedia*, i.e. the appearance of leg structures in the antennae. The treatments lead to cell death in the antennal rudiment; as a result proliferation increases in the rudiment, leading to 'transdetermination' analogous to that observed in pieces of rudiments cultured in vivo (p. 91). Conversely, the effect of the mutation can be 'corrected' by treating the rudiment with colchicine, which blocks mitotic activity.

Summarising, the confrontation of experimental data from the areas of molecular biology, genetics and embryology now yields a concrete picture of the automatism of development: it is no longer necessary to speculate to explain how the genes express themselves in the reproduction of structural patterns. Both transcription and the utilisation of its products are exclusively and rigorously controlled by cytoplasmic information. The precision with which cells of infinitely varied individualities 'work' in the organism results from the precision with which throughout development the summation of extracellular information of various sorts proceeds. We want once more to underscore an important point: the DNA contains neither a developmental message to be decoded, nor a developmental programme. It is comparable to the 'arithmetical and control circuits' of a computer (p. 24) in that it does nothing more than to process information provided by the 'input devices'. To think that a new gene must be activated each time a new structural feature emerges is tantamount to saying that one must make some change in the arithmetical and control circuits each time one submits a new calculation to a computer. A limited number of genes is theoretically sufficient to make an organism; making new species by evolution is theoretically possible without the addition of new nucleotide sequences.

The Programming of Development

The Composition of the Egg Developmental Programme

Mature, fertilisable eggs (oocytes) represent the final or terminal stage of differentiation of a specific cell lineage which originates in the 'primordial germ cells' of the embryo. Because these stem cells can also give rise to the male germ line (which terminates in the production of the spermatozoa) a choice must be made at some time between male and female. It is known that in vertebrates this choice is effected during organogenesis under the influence of the somatic (non-germinal) cells of the gonad. In both sexes the gametes are the products of two 'maturation' divisions. The first of these is the 'meiotic' division, the prophase of which is characterised by a particular appearance of the chromosomes, which carry numerous closely apposed side loops, each containing a core of unwound DNA that is actively transcribing. In amphibians they are called 'lampbrush chromosomes'.

In the female germ line the maturation divisions show several specific features. The nuclear activity needed for the production of reserve substances is particularly intense during meiotic prophase. The nucleus swells excessively and is then called the 'germinal vesicle'. This breaks down during meiotic metaphase, setting free a basophilic 'nucleoplasm' that mixes with the cytoplasm. Both the first (meiotic) and the second (mitotic) maturation division are highly unequal and take the form of the extrusion of a polar body. This happens in roughly the same area, the 'animal' pole region, opposite to the 'vegetal' pole region, towards which as a rule most of the yolk is accumulated. When the two maturation divisions are completed the egg centriole disappears. Meanwhile, at a stage of maturation that varies among animal groups, the egg enters a lethargic phase. Most biosynthetic processes and respiratory exchanges become suspended, until the entry of the sperm brings about the 'activation' of the egg. As a result protein synthesis and respiration are resumed and the maturation divisions, insofar as they were not completed, proceed. After the egg and sperm nuclei have fused (amphimixis) the process of 'cleavage' starts, i.e. the subdivision of the egg into blastomeres of progressively smaller size. The mitotic apparatuses (asters and spindles) controlling the cleavage divisions all descend from the sperm centriole: its first activity inside the egg is to form the 'spermaster'. Each cleavage plane appears at the equator of a cleavage spindle whose position in the cell is fixed by the development of the astral rays; these extend to the periphery of the cell and contribute to the formation of the 'cytoskeleton'. In the beginning of the cleavage period each nuclear division is succeeded directly by the prophase of the next division; there is no interphase and therefore no transcription of DNA. Consequently protein synthesis is directed by mRNAs produced, at least for the greater part, during oogenesis and stabilised by association with proteins ('masked messengers'). The first appearance of interphases does not occur at the same time in all blastomeres, which is reflected in a desynchronisation of the cleavage divisions.

Often at the beginning of maturation, but sometimes later, a process begins that is known as 'ooplasmic segregation'. The various organelles and other inclusions of the cytoplasm, also those that were until then more or less uniformly distributed, become displaced relative to one another. Thus, when the egg is ready to cleave it consists of different areas which are called 'ooplasms'. These can differ from each other in various respects such as pigmentation, ultrastructural configuration, accumulation of a particular type of inclusion (e.g. mitochondria, yolk platelets or ribosomes), or metabolic rate. Ooplasmic segregation ends with the ooplasms being sequestered into different blastomeres.

Although the egg was the object of the very first experiments aimed at the elucidation of the causes of development, and although it has served as a model cell for molecular biologists and its ultrastructure has been studied in minute detail, it has not definitively revealed the secret of the first beginnings of animal development. In fact, to synthesise this considerable body of knowledge would seem next to impossible at first sight. The eggs of the various animal groups differ widely in their morphology and properties. The cleavage stages do not lend themselves to the same experimental techniques, so much so that some experiments seem to have been motivated more by the wish to 'play' with various practical possibilities than by rigorously defined questions concerning the determination of the early rudiments. On top of all that, the interpretation of the experimental data is extremely tricky. The egg necessarily shares many activities with all other cells, but because its large size renders these activities more conspicuous they are sometimes taken as specific for the egg. For instance, we may take it that the surface waves observed to precede each cleavage in amphibian eggs are a common phenomenon in all dividing cells, only they will be much more difficult to visualise [*Hara* et al., 1980]. Then there is the matter of reserve substances. On the one hand, there is the whole machinery needed to synthesise the 'household' proteins during the first cleavage period, when the DNA is not transcribed due to the absence of interphases. On the other hand, there are the nutritional reserves needed to sustain the organism while it is still within its vitelline membrane. The dissociation of all these constituents from the molecules involved in the programming of development is made particularly difficult by the fact that they are also indispensable for the initiation of differentiation. This point is well illustrated by certain lethal mutations which lead to developmental arrest at a given stage but in effect do this by blocking the production of molecules that do not carry developmental information (such as rRNA in the case of the *nucleolusless* mutation in *Xenopus*).

Because of this very complex situation one cannot make proper use of the experimental data without having an approximate idea of the programming of development, which is in its turn derived from the ideas one has of the progression of differentiation in the embryo. If one realises that most authors still use the theory of sequential gene derepression – which has failed to provide a satisfactory solution for the problems of organogenesis – it is not surprising that the solutions proposed so far for the problem of programming do not tally with the total body of our present knowledge. This is the reason why, after having arrived at a new conception of the automatism of development in Part One, we must now look for other theoretical approaches to provide a new basis for the analysis of the observations and experimental results furnished by the literature.

Theoretical Approach to the Problem

> The egg contains in its cytoplasmic 'memory' the programme for the first few autonomous progressions, which result in the first heterotypic actions in development.

It is still said too often that development is programmed in the DNA because its mode and time schedule are specific for each species [*Ashby*, 1958;

Apter and Wolpert, 1965; *Apter,* 1966; *Arbib,* 1972; *Atlan,* 1972]. However, this view is formally contradicted by the results of many classical experiments. For instance, no amphibian somatic cell is capable of engaging in any kind of development, however abortive. Nevertheless, in certain species the nuclei of certain (particularly embryonic) tissues can sometimes support normal development when injected into enucleated eggs [*Gurdon,* 1964]. This shows that the DNA is 'manipulated' by the cytoplasm right from the start of development: *the egg is a differentiated cell just as any somatic cell whatever. Its capacity to give rise to a new individual rather than to participate in a particular tissue function is an indication that it has been specially instructed for this function.* This view is unambiguously supported by the fact that in certain lower animal groups, such as hydroids, planarians and slugs [*Laviolette,* 1954], eggs can originate from somatic cells that previously had participated in adult functions [review in *Nieuwkoop and Sutasurya,* 1981].

The cytoplasmic information present in the egg not only makes possible the start of development, it also determines its specific modalities. These are not always the same in the same species: in species which can reproduce both sexually (from an egg) and asexually (from the soma) the course taken by ontogenesis – and often even adult morphology – varies according to the origin of the cells from which it starts [*Brien,* 1968]. A case in point are the ascidians, where sexual reproduction involves a tadpole stage and the organisation of the tadpole is profoundly modified during metamorphosis, when the organism becomes sedentary. In contrast, during asexual reproduction the definitive structure is produced directly, without an intervening tadpole stage. The egg therefore contains, stored in its cytoplasmic 'memory', specific information that is used to establish the programme for development. The only condition for the execution of the programme is that the DNA belongs to the same species as the cytoplasm. This is apparent from the fact that hybrid fertilisation or interspecific nuclear transplantation always leads to developmental arrest at some stage [*Moore,* 1958], usually the gastrula stage.

As we have seen in Part One, the organisation of a cell population in the embryo involves the summation of the effects of extrinsic prepatterns, i.e. of programme components supplied successively to the population in question by neighbouring populations. As a consequence, at no time during development does a part of the organism contain the complete programme for its own further development. The younger a cell population, the more incomplete is the information stored in its 'collective memory' which it uses to organise itself. *A fortiori, no single part of the egg is capable of developing independently of the other parts beyond a certain, often very early stage* (this raises an apparent paradox when we think of the phenomena of embryonic regulation; however, this paradox will be resolved in the last chapter, p. 145). *It is the cytoplasm of the egg as a whole that contains a unique programme for subsequent cellular interactions:* (1) homotypic interactions which propel the cells through the autonomous progressions, and (2) heterotypic actions which re-initiate autonomous progressions, give them a definite direction, or block them temporarily. This programme we have previously called the *egg developmental programme* [*Chandebois*, 1980a].

To clarify the logical requirements for such a programme we will resort to a simile. Suppose we want to design a railway network on which mechanical trains travel without manual interference. We start with a single track that leaves a central station (the egg) and branches out to reach a number of peripheral end stations (the 'normal achievements' of the various prospective areas of the egg). In the central station a train will be placed on the track that consists of several carriages, each having a specific destination and representing a prospective area: each carriage must at some time be detached from the train to be shunted to the branch line leading to its proper end station (diversification of cellular activities). Each carriage is provided with its own clockwork. When the clockworks are released (fertilisation) the train will be propelled for a certain time (autonomous progression resulting from homotypic interactions). However, we may expect the train to come to a stop prematurely because its power runs out, in which case no carriages will be shunted off at the appropriate points (a situation comparable to the isolated prospective ectoderm forming a permanent blastula). To make the system work properly it will suffice to have a second train depart together with the first but on a separate track. Each train during its journey will trigger signals (heterotypic actions) commanding the manœuvres the other train will have to make: rewinding of the clockworks, shunts, and temporary arrests. If the clockworks are correctly set, reset and released, the system will work automatically: each carriage will reach its proper end station at the appointed time. Similarly, *for the production of the various tissue types to proceed automatically to the end, it is logically sufficient that the egg engenders two types of blastomeres which engage in two different and rigorously defined autonomous progressions.*

The only role of the first autonomous progressions resulting from the egg developmental programme is to provide the right conditions for the heterotypic actions. They can therefore be of short duration and do not have to produce any of the specific cell types of the embryo. For instance, the amphibian blastula consists of two 'moieties' [*Nieuwkoop*, 1969a] whose differentiation requires reciprocal inductions: if separated early they are incapable of organising themselves. The isolated 'ectodermal moiety' forms no mesoderm, neural plate or even epidermis. After producing embryonic ciliated cells distributed in a typical hexagonal pattern [*Grunz* et al., 1975], it dies. The isolated 'endodermal moiety' does not form endoderm but cytolyses almost immediately. This shows that in the endodermal moiety the role of autonomous progression is limited to calling into being (1) its inductive capacity, which will ensure the normal differentiation of the ectodermal moiety, and (2) its endodermal competence, which will only manifest itself under the influence of the ectodermal moiety after this has been partially transformed into mesoderm. Similarly, in the ectodermal moiety autonomous progression makes the cells pass through two successive phases of competence: mesodermal competence at the blastula stage, and neural and epidermal competence at the gastrula stage.

The most decisive heterotypic actions in organogenesis involve the transfer of information between cell populations originating from the different germ layers in the embryo. By way of example, let us recall that the whole ectoderm of vertebrate embryos organises itself under the influence of the mesoderm (e.g. neural plate) and the endoderm (e.g. pharyngeal clefts). Therefore, the disposition of the germ layers in the gastrula is a crucial pivot for the whole of development. If it succeeds, development automatically proceeds to its end (except when unexpected factors act on cell interactions). The normal course of gastrulation requires that a number of specific cell types are produced in the correct proportions (e.g. ectoderm, chordomesoderm, mesoderm and endoderm in the vertebrates). If there is for instance too much endoderm, it will fail to interiorise completely and the embryo will ultimately die as an exogastrula. Consequently, the problem of the programming of development can be reduced to that of the determination of the prospective areas for the ectoderm, mesoderm and endoderm – that which certain authors have called the 'programming of gastrulation'. The solution to this problem is not only to be sought in a knowledge of everything that happens before the egg starts cleaving; it requires first of all precise knowledge of what causes the transformations the various blastomeres undergo prior to gastrulation, the sort of knowledge that has been provided by the work of numerous experimental embryologists.

The Determination of the Germ Layers

The egg developmental programme consists of two components which act together in determining the germ layers in the gastrula. The first is the gene activation (GA) clock, which re-initiates DNA transcription. Set off at fertilisation, it provides two successive programmes of autonomous progression: first the 'endodermal' programme and then the 'ectodermal' programme; the latter in its turn provides the 'mesodermal' programme under the influence of an additional stimulus. The second component is the egg antecedent prepattern (EAPP), which is established during ooplasmic segregation. It allows part of the blastomeres to engage in the 'endodermal' programme.

According to the classical view, when transcription resumes during cleavage only a few genes would be activated; these would be different genes for each of the prospective areas of the ectoderm, mesoderm and endoderm. These areas would thus be determined by specific gene effectors localised there, which would be produced during oogenesis and are called 'cytoplasmic determinants'. Each type of determinant would at first be uniformly distributed in the immature oocyte and would later be sequestered in a particular area as a result of the cytoplasmic movements that lead to ooplasmic segregation. This would therefore imply a degree of 'preformation', i.e. a pre-existing organisation that would determine the localisation and extent of the anlagen of the various germ layers. In support of this concept one often cites the example of certain molluscs, particularly *Dentalium*. Here one has by electron microscopy demonstrated certain granules which seem to be determinants of the endomesoderm. They are located in the 'polar lobe', an extension of the egg near the vegetal pole. When they are displaced by centrifugation the embryo is arrested as an ectodermal vesicle, just as is the case after removal of the polar lobe [*Dohmen and Verdonk,* 1979]. However, this is an exceptional case and we should not make a principle out of it. In general a prospective area is determined by the local accumulation of mitochondria or yolk platelets, which are also found elsewhere throughout the egg, but in lower amounts [*Dalcq,* 1961]. And the more we study the details of the available experimental data, the more we find that the 'classical' views unreservedly adopted by molecular embryologists are divorced from factual reality.

It has long been known that the determination of the future germ layers does not occur in the same way in different animal groups. In some groups, particularly the ascidians, it is effectively completed for *all* the future germ layers when the egg starts cleaving. Any part of the egg, if isolated, forms the same structures it would form until the gastrula stage (i.e. its 'normal achievements' for that stage). These are the groups with so-called *strictly mosaic eggs.* In other groups (the majority of the animal kingdom) the determination of at least one germ layer results from induction (the prototypes here are the sea urchins and the amphibians). Consequently a part of the egg, when isolated, may at the time of gastrulation perform the achievements of another part; this is characteristic of the so-called *regulative eggs.*

We should emphasise right away that this diversity in egg types has nothing to do with evolution. In the one group of the higher insects we find both mosaic and regulative eggs [references in *Counce,* 1973]. Certain species are so highly determinative in their development that the rudiment of an imaginal organ can be destroyed in the egg a few hours after fertilisation without damaging the homologous larval rudiment. Other species exhibit extreme regulative capacities, as exemplified by no less than 100 embryos originating from a single egg, or by the germ band being reconstituted repeatedly from the extra-embryonic ectoderm (so-called successive polyembryony). Despite the diverse nature of the experimental data it is not very likely that the principles of developmental programming would differ so radically from one animal group to the other. It is more plausible to assume that the principles of programming are universal without excluding great flexibility from one form to another. Then this very diversity can but simplify the analysis of a problem that would otherwise appear inextricable: it should enable us to eliminate those events in the course of the earliest phases of development that are only of secondary importance.

The key to the problem is provided by the embryogenesis of the mammals, because it is at once most aberrant and best known. In this group the external blastomeres of the morula form a superficial cell layer which represents the trophectoderm (the future trophoblast) while the remaining blastomeres constitute the 'inner cell mass' (the embryonic knob). This detaches itself partially from the trophectoderm, leading to the appearance of the blastocoel. The cells of the inner cell mass facing this cavity constitute the future extra-embryonic endoderm. Experiments have shown beyond doubt that at the beginning the cells are all alike. But they start diversifying soon because their environment differs, so that they receive different extracellular information [*Graham,* 1971; *Gardner and Rossant,* 1976; *Graham and Kelly,* 1977].

The cells of the trophectoderm are the first to be determined as a result of their peripheral location in the morula [*Hillman* et al., 1972]. Next the endodermal cells are determined by their location on the surface of the inner cell mass facing the blastocoel [*Tarkowski and Wroblewska,* 1967; *Rossant,* 1975]. Finally, inside the inner cell mass, the mesodermal cells are determined by an inductive influence emanating from the endoderm [*Levak-Svajger and Svajger,* 1970]. It is known that transcription of the DNA is resumed as early as the 2-cell stage, when interphases appear [*Woodland and Graham,* 1969]. Consequently, at that stage all blastomeres appear to be engaged in the same autonomous progression, in which the nuclei participate. In the course of this autonomous progression the cells pass through three separate periods of competence: trophectodermal, endodermal and mesodermal. If they do not receive the extracellular information required for the manifestation of any of these competences they develop into embryonic ectoderm. In other words, *at two stages in a single sequence of transformations of their cytoplasmic information content certain cells are shunted into an 'endodermal' programme, others into an ectodermal programme, while part of the latter cells are finally shunted into a 'mesodermal' programme by an inductive influence emanating from the cells which are beginning to execute the 'endodermal' programme* (fig. 18c).

In oviparous species which have regulative eggs, embryonic cells that are developing in ectodermal direction also require an inductive event to form mesoderm. The induction emanates without exception from the vegetative parts of the embryo, made up by cells that belong to the endodermal germ layer, even if they do not always participate in the formation of the digestive tract. We already know the experimental work that has shown this in the amphibians (p. 35). In birds the mesoderm is formed from the epiblast (which may be regarded as analogous to the 'ectodermal moiety' of the amphibian embryo): it develops entirely into ectoderm if the hypoblast (the 'endodermal moiety'), which forms the embryonic endoderm, is removed [*Eyal-Giladi,* 1970]. In sea urchins the most vegetal cells of the embryo are the small micromeres. They are situated in the centre of the prospective area of the archenteron, but detach themselves from it prior to gastrulation to form mesenchyme. If they are grafted to the prospective ectodermal area they convert the surrounding cells to endo-mesoderm, so that a second archenteron forms [*Hörstadius,* 1935]. In certain insects the cells of the posterior pole of the embryo produce substances which are necessary for the formation of the germ band on the ventral surface of the blastoderm: if one isolates them by a tight ligature only extra-embryonic ectoderm is formed [*Seidel,* 1929]. In

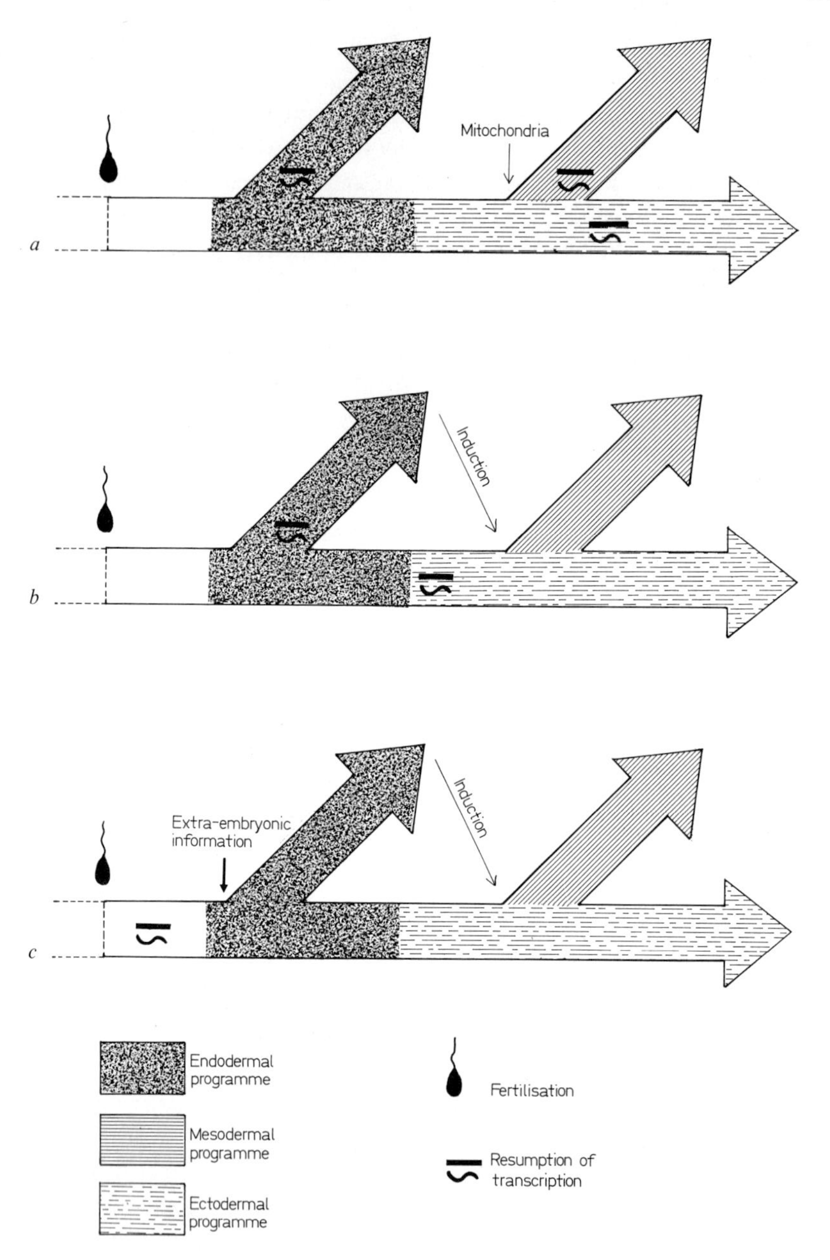

Fig. 18. Analogies and differences in the way in which blastomeres are shunted into the 'ectodermal', 'endodermal' and 'mesodermal' programmes. *a* Annelids, molluscs and ascidians. *b* Sea urchins and amphibians. *c* Mammals. Explanation in text.

certain insects it has been established that part of the cells of the posterior pole of the embryo take part in the formation of the midgut [references in *Anderson*, 1973]. *Therefore, as in the mammals, part of the blastomeres are engaged in an 'endodermal' programme, the other blastomeres in an 'ectodermal' programme. The latter comprises a 'mesodermal' programme, into which are shunted only those blastomeres that come under the inductive influence emanating from the cells which are executing the 'endodermal' programme.*

In all animal groups with regulative eggs except the mammals the acquisition of inductive capacity by the 'vegetal' cells appears to be linked up with the re-initiation of transcription in their nuclei, which occurs earlier than in the other blastomeres. *If this early transcription is prevented or just retarded the differentiation of the cells is ectodermal in character.* In the sea urchins, for example, the first interphase nuclei appear in the micromeres at the vegetal pole between the 4th and the 5th cleavage. It is known that the inductive property of the micromeres is linked up with the activity of their nuclei [*Hörstadius*, 1973] and probably with the production of a specific mRNA [*Czihak*, 1965]. While the micromeres form mesenchyme upon isolation, they show a tendency towards ectodermal differentiation when they are enucleated [*Lorch* et al., 1953]. When the egg is treated with certain substances such as lauryl sulphate the cleavages become equalized: the cells of the vegetal pole area cleave in the same rhythm as the other blastomeres. They do not take on the appearance of micromeres, form neither mesenchyme nor spicules, and fail to induce an archenteron: the embryo is arrested as a permanent blastula [*Tanaka*, 1976]. In the amphibians cleavage first slows down in the large cells of the vegetal yolk mass. This desynchronisation indicates the resumption of RNA synthesis and coincides with the start of mesoderm induction [*Signoret*, 1980; *Kirschner* et al., 1980]. If the egg is kept in an inverted position so that the coarse yolk accumulates towards the animal pole, the cleavage pattern also becomes inverted [*Stanisstreet* et al., 1980] and the cells of the original vegetal region form ectoderm [*Pasteels*, 1938]. Similarly, in the insects [references in *Counce*, 1973] the first interphases appear in the cells of the posterior pole, which owe their specific properties to the resumption of transcription. If the colonisation of the posterior part of the embryo by cleavage nuclei is prevented the effect is that of a ligature at a later stage: the germ band does not form [*Seidel*, 1929]. All these facts strongly suggest that in all these species with regulative eggs *the appearance of an 'endodermal' and an 'ectodermal' programme does not result from the activities of two qualitatively distinct areas of cytoplasm. It represents two stages in one and the same sequence of transformations undergone by cytoplasm that is totipotent from the start* (fig. 18b), *as is the case in mammals. However, unlike in mammals obviously the nuclei are not involved in the production of these successive programmes. Evidently a molecular machinery is set going in the cytoplasm upon activation and is already functioning before transcription of the DNA is resumed. The 'endodermal' programme, which is of only limited duration, is 'picked up' or 'intercepted' by the first cells in which mitotic interphases appear. If it is not intercepted at the proper time the cells will 'pick up' the 'ectodermal' programme.*

The mosaic eggs appear at first sight to have entirely different properties. The fates of the prospective areas for the three germ layers are fixed before the first mitotic interphases appear. For instance, in ascidians, where the cleavage divisions remain synchronous until the 5th cleavage [*Reverberi and Minganti,* 1946b], the prospective ectoderm, endoderm and mesoderm as early as the 3rd to 4th cleavage already contain the information necessary to perform their achievements when they are isolated [*Reverberi and Minganti,* 1946a]. In various groups, such as the annelids and ascidians, the prospective areas of the three germ layers only differ in the relative concentrations of common organelles and cytoplasmic inclusions. Mitochondria are very numerous in the prospective mesoderm and a little less numerous in the prospective ectoderm, where the endoplasmic reticulum is predominant. They are virtually absent in the prospective endoderm, where the greater part of the yolk is concentrated [*Lehmann,* 1958; *Reverberi,* 1961; *Devriès,* 1973a]. It has become clear that the distribution of the mitochondria is the determining factor in the localisation of the prospective areas. In the ascidians the muscle lineage originates from the area where they accumulate, either as a result of ooplasmic segregation during normal development or when they are redistributed by centrifugation of the unfertilised egg [references in *Reverberi,* 1961]. Although it has been assumed that the mitochondria play a specific role in the determination of the muscle lineage [*Reverberi,* 1972], many authors now reject this idea [references in *Dohmen and Verdonk,* 1979]. It has been shown that their activity only relates to that of the ground cytoplasm [*Devriès,* 1973a; *Whittaker,* 1979], which is totipotent throughout [*Ortolani,* 1958].

This last point is of paramount importance because it suggests that the same molecular machinery can produce the various different programmes of the first autonomous progressions independently of any nuclear activity. If this machinery would not differ fundamentally from that operating in regulative eggs, it would first produce the 'endodermal' programme, then the 'ectodermal' programme, the 'mesodermal' programme deriving from the latter as a result of an additional stimulus. In this view only a slowing-down of metabolic activity in a given part of the egg could assure that the 'endodermal' programme is still available there, while the rest of the egg has already produced the 'ectodermal' programme (fig. 18a). This prediction tallies very well with the experimental data. In ascidians blastomere isolation and recombination experiments have shown that notochord, muscles and ectoderm are already stably determined at a stage when the prospective endoderm is still capable of forming mesoderm [*Tung* et al., 1941]. We know that the mitochondria are the suppliers of energy: the more numerous they are, the higher will be the level of biosynthesis. Now, it is precisely in the area that has become devoid of mitochondria as a result of ooplasmic segregation that the blastomeres will be shunted into the 'endodermal' programme. It is remarkable that in mosaic eggs a high density of mitochondria has the same effect as an inductive event in regulative eggs: the selection of a 'mesodermal' programme from the 'ectodermal' one.

The two mechanisms through which the temporarily available 'endodermal' programme can be intercepted (i.e. desynchronisation of mitoses or local retardation of metabolic activities) are not mutually exclusive. It is indeed probable that the two are used jointly, though in varying proportions, which would explain the variety of properties exhibited by various animal eggs upon experimental interference. In an annelid species both the distribution of the mitochondria and the time of incipient transcription play a determining role in the formation of the prospective ectoderm [*Devriès,* 1973b, 1976]. In sea urchins only mitotic desynchronisation seems to be involved, since the mitochondria are uniformly distributed. Nevertheless, in the animal region, i.e. the prospective ectoderm, anabolic metabolism and the reducing activity it entails is higher. Treatment with lithium ions, which reduces anabolic metabolism, leads to overdevelopment of the endoderm. On the other hand, stimulation of oxydation favours animalisation, i.e. the formation of ectoderm [references in *Lallier,* 1964; *Gustafson,* 1965]. This co-existence of the two mechanisms of interception could explain the paradoxical effects of certain treatments. For instance, in sea urchins the period of synchronous cleavages is prolonged by the effect of lithium [*Dettlaff,* 1964]; this would have to result in overdevelopment of the ectoderm, were it not that lithium at the same time reduces metabolism in the animal region.

Whether eggs predominantly use one or the other of the two mechanisms to 'catch' the 'endodermal' programme is probably not a matter of chance. It is well known that oocytes do not become fertilisable at the same age. In one group of animals, which comprises the most typical regulative eggs, the

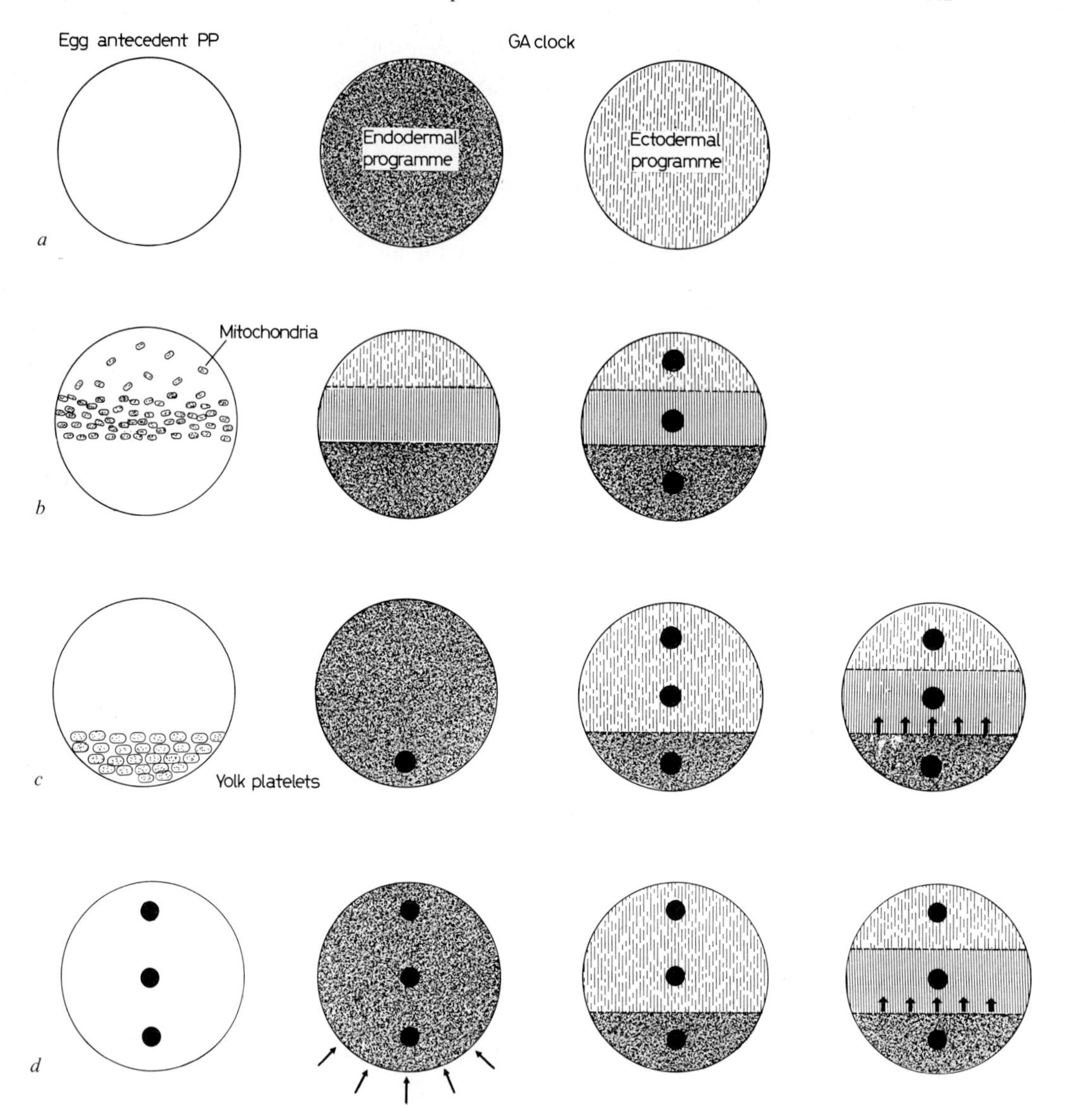

Fig. 19. Interference of the gene activation (GA) clock and the egg antecedent prepattern (EAPP). Full circles, interphase nuclei; other symbols as in figure 18. *a* In the absence of an EAPP, the GA clock provides first the 'endodermal', then the 'ectodermal' programme including the 'mesodermal' programme. *b* The EAPP desynchronises the cytoplasmic activities in the various parts of the egg previous to cleavage. When the nuclei enter mitotic interphase the three programmes are already available (annelids, molluscs, ascidians). *c* The EAPP desynchronises the cleavage mitoses. The first nuclei to enter mitotic interphase intercept the 'endodermal' pro-

sperm is accepted during the second maturation division (vertebrates) or even after the extrusion of the second polar body (sea urchins). Here activation occurs shortly before the start of cleavage, and cleavage is already far advanced when the 'endodermal' programme is made available. It can therefore be 'caught' through desynchronisation of cleavage divisions. In contrast, in the other group of animals, to which belong the most typical mosaic eggs, the sperm is accepted early, i.e. during (ascidians) or even before (annelids) the first maturation division. Because the maturation divisions must be completed before amphimixis can occur, cleavage begins only long after activation. If part of the cytoplasm were not devoid of mitochondria, when transcription starts all blastomeres would possess the 'ectodermal' programme and none the 'endodermal' one. The early determination of the prospective germ layers in mosaic eggs is usually attributed to an acceleration of biosynthetic processes (expressed by the term 'tachygenesis'), but although this cannot be completely excluded a priori the early time of activation is probably a more important factor.

In summary, then, *the transformation of the unfertilised egg into a gastrula requires the cooperation of two types of organisation: one in the dimension of time and the other in the three dimensions of space* (fig. 19) [*Chandebois*, 1980a]. Therefore, the principle of the programming of gastrulation is akin to that of the programming of a readjustment in organogenesis, during which a structural pattern is reproduced through the cooperation of an autonomous progression (progressive change of cytoplasmic information content) and a previously acquired spatial organisation (the antecedent prepattern). We have called the molecular machinery which is set going at fertilisation and successively produces the 'endodermal' and the 'ectodermal' programme the 'gene activation clock' (GA clock). The features of the egg's organisation which enable the blastomeres to be shunted into one of the two programmes, thus determining the localisation and extent of the prospective areas of the germ layers, constitute the 'egg antecedent prepattern' (EAPP). The programming of development in mammals appears to be a special case in that it does not involve the establishment of an EAPP. This does not invalidate

---

gramme. The nuclei which enter interphase later intercept the 'ectodermal' programme. Some of these cells are finally shunted into the 'mesodermal' programme, as a result of induction (arrows) emanating from the endodermal area (sea urchins, amphibians). *d* The egg does not use an EAPP and interphases appear right from the start of cleavage. Extra-embryonic information (long arrows) is required for part of the cells to be shunted into the 'endodermal' programme (mammals).

the comparison with a readjustment, however. In fact in this group the extra-embryonic milieu provides an *environmental prepattern* to the egg: its presence is sufficient to ensure the differentiation of the trophectoderm, the formation of the blastocoel and the determination of the endoderm.

Finally we want to stress that the execution of the egg developmental programme requires a certain concordance between the course of cleavage on the one hand, and the functioning of the GA clock and the establishment of the EAPP on the other. In mosaic eggs the positioning of the cleavage planes must be rigorously fixed to ensure that local concentrations of cytoplasmic inclusions occur in the proper blastomeres. In regulative eggs the time course of cleavage must be precise to ensure that part of the blastomeres are shunted into the 'endodermal' programme. Therefore, in the next chapter, which deals with the two components of the egg developmental programme, we will also consider the factors that determine cleavage.

The Establishment of the Egg Developmental Programme

The Gene Activation Clock

The cytoplasm of the egg contains the stabilised 'primers' needed for the reactivation of all structural genes. Upon activation of the egg the first primers to be destabilised are those that furnish the 'endodermal' programme. The other primers are destabilised later and furnish the 'ectodermal' programme. The 'mesodermal' programme derives from the latter under the influence of an inductive event or a local accumulation of mitochondria. The synthesis of primers for all structural genes accounts for the generalised transcription of the DNA during meiotic prophase and for the presence of tissue-specific proteins in the oocyte. The reactivation of transcription implies (1) the continuous presence, from one generation to the next, of a particular class of primers specific for the germ line, and (2) the registration of extracellular information.

The production of the two differentiation programmes by the egg cytoplasm cannot be discussed without having a precise idea of how transcription is re-initiated during cleavage. Unfortunately, investigations in this field have only provided an incomplete inventory of substances occurring before and after fertilisation, and some disparate particulars of their synthesis and the roles they perform. They have not yet resulted in a coherent view and can only provide arguments for or against concepts developed on other grounds. For the problem that concerns us here, the proper use of these arguments will be extremely useful, for we are still faced with a choice between two theories. The theory based on sequential gene derepression is now accepted as classical but it cannot explain the majority of the biochemical features of the egg, a drawback that comes on top of its failure to provide a correct explanation of embryonic induction (see Part One). In contrast, the theory of differentiation which we have proposed provides meaning to certain activities of the oocyte that so far had remained enigmatic. What is more, it allows us to *predict* the existence of those activities while searching for the requirements the theory makes for the programming of development. It is from this point of view, which is theoretically the most interesting, that we will now reconsider the various data available in the literature.

Our novel concept was based on the idea that a structural gene sustains its own activity by using as a 'primer' the first portion (the Hn1RNA) of the HnRNA that is transcribed from it. *In fact many authors now subscribe to the notion that cytoplasmic determinants are particular RNAs produced during oogenesis and stabilised by association with proteins.* We will mention two of the most convincing studies. In dipteran insects a mutation occurs as a result of which the egg gives rise to an embryo with two posterior ends (the 'bicaudal' syndrome [*Bull,* 1966]). One has been able to mimick this anomaly experimentally by applying to the anterior end of normal eggs, before its colonisation by nuclei, treatments which degrade RNA (UV [*Kalthoff,* 1971] or ribonuclease [*Kandler-Singer and Kalthoff,* 1976]). The treatment obviously destroys the RNAs that function as cytoplasmic determinants, and for some reason as yet unknown these are replaced by posterior determinants which were apparently stabilised at the time of treatment. In ascidians the injection of whole RNA extracted from ovaries into animal blastomeres at the 8-cell stage causes them to form neural structures of posterior type which are normally produced by vegetal blastomeres [*Ortolani and Marino,* 1973]. The primers in question apparently do not belong to the class of mRNAs but to another class of RNAs [*Whittaker,* 1979]. It seems not impossible that we have to do with Hn1RNA. In fact it has been shown in amphibians that the HnRNA molecules, which are produced during meiotic prophase and which are very long, pass into the cytoplasm at the time of germinal vesicle breakdown. The RNA molecules resulting from their processing are stabilised, but they do not all represent mRNA [*Sommerville,* 1979]. More recently *Thomas* et al. [1981], on the basis of a review of numerous papers, have pointed out that the greater part of the RNA present in the cytoplasm of the mature egg – which is unprocessed RNA – shows an organisation comparable to that of the nuclear RNA of somatic cells. In addition to message sequences identical to those found in maternal mRNA, it contains sequences that are not translated, among them 5′ leaders. The authors ascribe an important function to these additional sequences. They could be taken up by the blastomere nuclei and could then at the same time initiate and programme gene expression in them.

We have seen that a gene seems to be definitively repressed when its transcription products disappear from the cells concerned (p. 18). In this view *the capacity of the oocyte to give rise to all the various cell types of the organism requires that all primers needed for the reactivation of the whole 'structural' genome be stored in the oocyte in numerous copies. Because the synthesis of a primer cannot be dissociated from that of the adjacent messenger, during oogenesis the whole genome must be transcribed repeatedly. As a result we may expect to find in the oocyte all the mRNA classes of the organism, particularly those coding for the luxury proteins of the functional tissues. And even the presence of these luxury proteins themselves in the oocyte may be predicted, at least insofar as the translation of their messengers is possible.* And in effect there seems to be no repression whatever during meiotic prophase. Each of the innumerable loops which characterise the lampbrush chromosomes represents a coding sequence unit in the DNA engaged in transcription. It is difficult to explain this extraordinary activity from the need to store proteins required for the nutrition of the early embryo, and some authors [*Sommerville*, 1977; *Perlman* et al., 1977] have already foreseen its possible role in the programming of development, without however being able to specify it further.

From the moment the germinal vesicle has broken down the egg contains in stored form all the information needed for development [*Smith* et al., 1968]. The amount of information contained in the mRNA class of molecules is considerable: at least 40,000 times that contained in the gene that codes for β-haemoglobin [*Davidson and Hough*, 1971]. The pool of HnRNA does indeed contain sequences that are involved in luxury metabolic strategies of the later functional tissues. For example, the messenger for myosin has been identified in the egg [*Dym* et al., 1979]. The unexpected presence in the egg of proteins specific for adult tissues has now been confirmed unambiguously. So far it was attributed to accidental and negligible phenomena such as 'leaky' genes [*Sommerville*, 1977]. One has so far identified larval and adult globins [*Perlman* et al., 1977], a lens-specific crystallin [*Flickinger and Stone*, 1960], chondroitin sulphate [*Kosher and Searls*, 1973], as well as ecdysterones in insects [*Hsiao and Hsiao*, 1979].

The re-initiation of transcription during cleavage implies that the primers stored in the cytoplasm must be taken up by the nuclei, possibly still complexed with mRNA and the corresponding protein. After being destabilised the primers must then bind to the DNA, each of them at a site analogous to the one from which it was originally transcribed. It has indeed been demonstrated that all information encoded in RNA 'inherited' from the egg (i.e. wholly maternal information) is present until the mid-blastula stage. The first signs of its utilisation are observed when the activity of the genome increases appreciably [*Davidson*, 1967]. Although the RNA concerned is of cytoplasmic origin, it enters the nuclei during cleavage [*Davidson and Hough*, 1969]. In that period the nuclei also accumulate proteins produced during maturation [*Ecker and Smith*, 1971].

As in the oocyte, in the cleaving egg as a whole the re-initiation of transcription must necessarily concern the whole genome. And in fact no essential changes have been found to occur in the protein pool after transcription is resumed. Neither does a new important mRNA class appear, nor do transcription products disappear that were already abundant in the fertilised egg [*Infante and Heilmann*, 1981]. Apart from this, the resumption of nuclear activity is attended by the appearance of proteins of 'paternal' or 'nuclear' type, as a result of the initiation of transcription of genes introduced with the spermatozoon. For certain of these it has been shown that their 'maternal' homologues are already present in the cytoplasm of the oocyte. We may mention several enzymes in the frog [*Wright and Subtelny*, 1971] and an endoderm-specific phosphatase in ascidians [*Whittaker*, 1979].

This concept of how the genome is reactivated, summarised diagrammatically in figure 20, provides an explanation for experimental results that so far seemed difficult to interpret: for instance, the maternal-effect mutation *o* in the axolotl [*Brothers*, 1976]. The eggs of females homozygous for this mutation *(o/o)* cleave but the embryos fail to differentiate, whatever the genotype of the father. As regards RNA and protein synthesis the eggs behave as if they had been enucleated. The blastomeres cannot be 'rescued' by grafting them to a normal embryo and do not participate in its development. The eggs, however, can be 'rescued' by injection of material derived from wild-type eggs and embryos; the compartment from which this material can be taken varies with the stage of development. The compartments which give positive results are precisely those where all the primers for gene reactivation would be expected to be located (fig. 20). They are on the one hand the germinal vesicle and the nuclei of the late blastula, where all the DNA is being transcribed and therefore produces all primers, and on the other hand the cytoplasm of mature oocytes and eggs in the early stages of cleavage, where the stabilised primers are stored. In contrast, extracts from early blastula nuclei, which have not yet been reactivated – and therefore do not yet produce primers – are ineffective. *Brothers* [1976] thinks that the mutant fails to produce a protein that is indispensable for gene reactivation, but to us it seems more likely that the effect of the mutation involves a disturbance leading to the premature degradation of the primers.

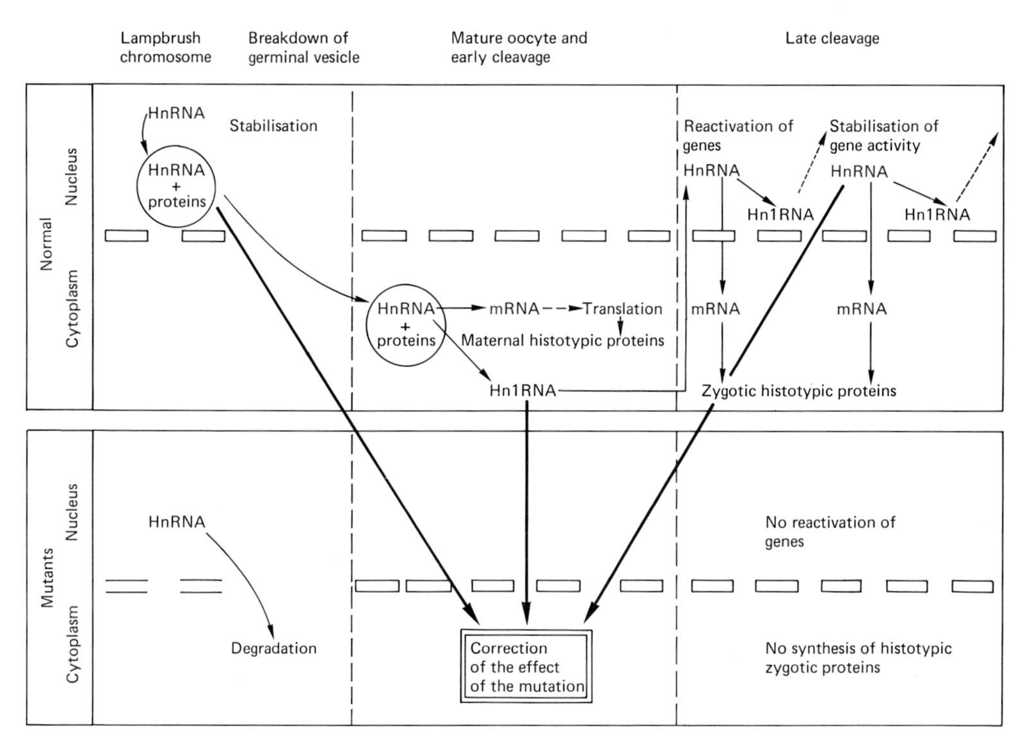

Fig. 20. Principles of gene reactivation in early development. Above: Synthesis, stabilisation and utilisation of primers (Hn1RNA) in normal development. Below: Absence of transcription in mutant eggs in which the primers are not stabilised (or prematurely destabilised) and consequently prematurely degraded. The effects of such mutations can be corrected by injection of material taken from wild-type oocytes, uncleaved and cleaving eggs (arrows). Explanation in text.

Because experiments have shown that the 'endodermal' programme is available before the 'ectodermal' one, we must assume that the primers are not all destabilised at the same time because of differences in their constitution – and perhaps also because of physico-chemical changes occurring in the cytoplasm after fertilisation. Thus, the primers for the genes involved in the specific activities of the tissues to be formed by the 'endodermal' part of the egg would be the first to become available. If at this moment these primers would not find themselves close to interphase nuclei, they would be degraded and would have disappeared more or less entirely when the other primers are destabilised in their turn (fig. 21). This general scheme is not sufficient, however. We still have to explain why the determination of the same prospec-

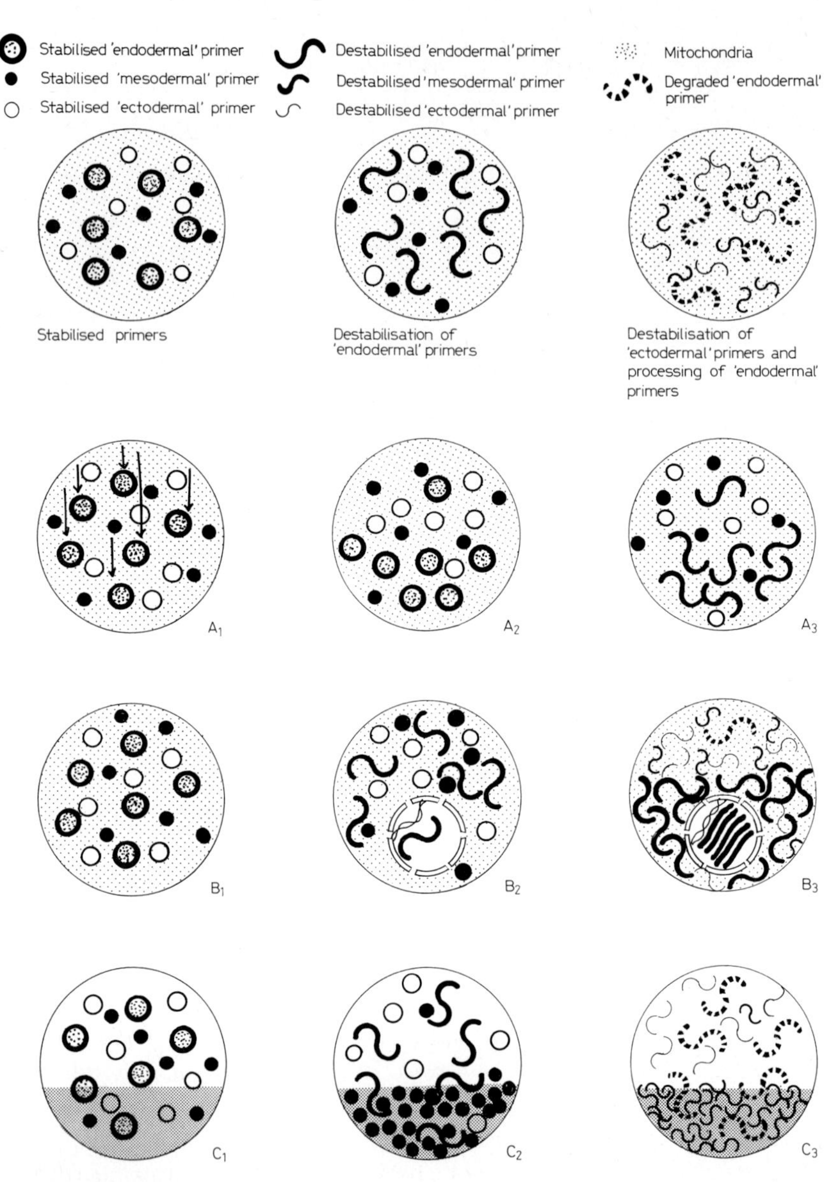

Fig. 21. Mechanisms involved in the interception of the 'endodermal', 'ectodermal' and 'mesodermal' programmes. Above: The successive production of the programmes is due to a time shift in the destabilisation of the primers. Below: Three possible procedures by which a local concentration of primers can be obtained that is sufficient to effect determination: A, relocation of large-size primers; B, re-initiation of transcription; C, local accumulation of mitochondria allowing for the replication of primers independent of nuclear activity.

tive area does not occur in the same manner in all animal groups. For instance, why does the muscle lineage of the prochordates appear there where mitochondria accumulate prior to fertilisation, whereas in the vertebrates, so closely related to the prochordates, it originates from cells that come under the inductive influence of the 'endodermal moiety'? Why is extra-embryonic information required in the mammals for the determination of the 'endodermal moiety'? Why is it that in certain molluscs the vegetal area is determined by RNA granules, which are not found in other groups? We also have to explain why a prospective area whose fate seems to be fixed can sometimes be made to perform achievements of another area until a more or less advanced stage. Incidentally, we must stress here that the theory of sequential gene derepression has no answer for all these questions.

In the determination of any given organ rudiment only the result counts, i.e. certain specific substances (as well as the primers of the genes coding for those substances) must exceed a certain concentration threshold (p. 42). The additional presence of other latent metabolic strategies does not influence this result; it only confers on the rudiment certain potencies which may manifest themselves under the proper experimental conditions. Therefore, we should not think that the determination of a prospective area is due to the *exclusive* presence of a particular class of destabilised primers. What is important is that those primers reach the required concentration, and this can be effected through various different means (fig. 21). First, primers can be located to their proper prospective area through ooplasmic segregation (fig. 21, A_{1-3}). This requires them to be part of granules of a certain size, which seems to be so in exceptional cases only (we have mentioned the granules in the polar lobe of certain molluscs). Second, when transcription is resumed in the nuclei when a certain class of primers is available in the cytoplasm, their concentration can be enhanced by the production of new HnRNA (fig. 21, B_{1-3}). Just as we have seen for determinative events later in development, positional information may be required to raise the rate of transcription, for instance induction in the case of the mesoderm in regulative embryos, or extra-embryonic information in the case of the mammalian endoderm. Finally, in the ascidian egg granules that have been taken to be the 'determinants' of the muscle lineage gradually form from the ground cytoplasm, there where mitochondria are accumulated [*Conklin*, 1931]. One is tempted to imagine that in this case the primers could be produced in the cytoplasm (fig. 21, C_{1-3}), a hypothesis suggested by the observation that cytoplasmic DNA is transcribed after fertilisation [*Brachet*, 1974].

This concept of determination of the first prospective areas tallies with the biochemical data: at the start of cleavage spectacular changes occur in the concentrations of many different abundant RNA classes which do not code for substances involved in the common functions and metabolism of all cells (household substances) [*Infante and Heilmann*, 1981].

The transcription of the whole 'structural' genome at the lampbrush chromosome stage remains the pivot of the programming of development. We must now enquire how this in its turn is determined; even though we know nothing of the nature and role of the substances that re-initiate it [*Smith and Williams*, 1979].

Because the loss of specific primers in somatic cells means that the corresponding genes in them can no longer be reactivated, it would be tempting to assume that in the female 'germ line', from the egg of one generation to that of the next, the synthesis of all primers continues without interruption. Not only is nothing known that might support this idea, but we now know that at least in mammals the germ cells temporarily lose their totipotency: as in somatic cells, the whole paternally derived chromosome becomes heterochromatic, and consequently transcription on it ceases. It is only re-established at meiotic prophase [*Monk*, 1981]. This suggests that the resumption of transcription requires the presence of a non-specific primer which functions as a sort of 'master key'. We know that the 5′ end of HnRNA (which is the first to be transcribed) contains repetitive sequences which are similar in all operons. Therefore, it could be suggested that a species of RNA is stored in the oocyte which is shorter than the specific primers and is complementary to a sequence of nucleotides in the DNA that is shared by all initiation sites [*Chandebois*, 1981].

Whatever the nature and quantity of the primers which re-initiate the generalised transcription of the DNA, their synthesis is the result of the specific activity of the germ cells (it should be remembered that the appearance of the chromosomes during meiotic prophase is virtually the same in sperm and oocyte). In certain species the germ cells are characterised by specific inclusions which contain proteins and are called 'germinal granules'. They have been encountered at all stages during the development of the germ line, although they have been found to disappear temporarily before the oocyte stage in the Diptera [*Mahowald* et al., 1979 a] and in the urodele amphibians [*Ikenishi and Nieuwkoop*, 1978]. Unfortunately, we still do not know which role they play in the initiation of meiosis.

The proteins contained in the germinal granules are obviously analogous to the products of the dominant metabolic strategy of a differentiated tissue. This becomes apparent from electron microscopic studies on *Hydra* [*Noda and Kanai,* 1977]. In these animals germ cells and somatic cells are constantly engendered by the same type of generative cell: the interstitial cells. These cells contain typical germinal granules, which disappear when the cells develop into somatic cells but become more numerous when they become gametes. An analogous observation has been made on the pole cells of *Drosophila,* special blastomeres at the posterior pole of the egg which contain germinal granules. The pole cells do not all form germ cells; about half of them take part in the formation of the midgut, and in these cells the germinal granules disappear altogether [*Mahowald* et al., 1979 a]. This already suggests that *the germ line is effectively determined when the amount of germinal granules exceeds a given threshold.*

In various animal groups, particularly the amphibians and dipterans, the germinal granules are already seen in the egg before cleavage begins. But at that time they contain RNA [*Blackler,* 1958], which has the characteristics of mRNA [*Mahowald and Hennen,* 1971] of maternal origin [*Lundquist and Emanuelson,* 1980], i.e. produced in the oocyte. This disappears towards the end of cleavage: in the anurans at the gastrula stage [*Blackler,* 1958], when the granules come to lie closer to the nucleus [*Smith and Williams,* 1979], and in the dipterans at the blastoderm stage. Obviously the germinal granules of the egg form part of the store of 'primer'-mRNA-protein complexes produced in the oocyte at the start of meiosis. Experimental data show that they in fact behave in the same manner as certain germ layer determinants. They accumulate in the so-called 'germinal cytoplasm' situated near the vegetal or posterior pole of the egg (in amphibians this is also called the 'subcortical cytoplasm' and in *Drosophila* the 'polar cytoplasm'). The blastomeres in which this cytoplasm ultimately comes to lie give rise to primordial germ cells. When the polar cytoplasm of an unfertilised or fertilised *Drosophila* egg is injected into the anterior part of another egg at the start of cleavage, this results in the heterotopic formation of germ cells capable of developing into gametes [*Illmensee and Mahowald,* 1974].

Again in *Drosophila,* the displacement of the germinal granules by centrifugation leads to sterile adults [*Jazdowska-Zagrodzinska,* 1966]. Moreover, two mutations are known in the fruit fly which bring about sterility of the adults: *grandchildless* [*Mahowald* et al., 1979 b] and *female sterile Nas* [*Kern,* 1976]. In the eggs of these mutants the germinal granules disappear 'spontaneously', i.e. without lysosomal activity being involved. It appears, therefore, that in normal eggs the activity of a gene (or genes) is required for their stabilisation. If this fails to occur their degradation is advanced and the reactivation of the genes involved in the dominant metabolic strategy of the germ cells apparently becomes impossible. It should be noted that these two mutations seem to involve a deficiency analogous to that produced by the *o* mutation in the axolotl (p. 118, fig. 20); their effects would differ only because they affect the primers of different genes.

The activity of the germinal granules is apparently linked up with the RNA present in them. In effect, in anurans UV-irradiation of the vegetal pole of the egg leads to the absence of germinal granules [*Bounoure,* 1937], but this effect can be counteracted by injecting subcortical cytoplasm into the irradiated region [*Smith,* 1966]. In *Drosophila* UV-irradiation of the posterior pole of the egg also leads to sterile adults [*Graziosi and Micali,* 1974]. Other experiments in the fruit fly suggest that germinal granule material returns from the cytoplasm to the nucleus (which reminds us of the re-entry of somatic cytoplasmic determinants into the nucleus prior to the resumption of transcription, p. 118, fig. 20). Among the interphase chromosomes of early pole cells so-called 'nuclear bodies' were observed which were ultrastructurally analogous to germinal granules. They originate from the cytoplasm, for if one injects germinal granules from *Drosophila immigrans* into a *melanogaster* egg, the nuclear bodies are of the donor type [*Mahowald* et al., 1976].

In conclusion, then, there is every reason to think that the *germinal granules of the egg contain one or more primers which are indispensable for the re-initiation, at the start of cleavage, of the transcription of one or more genes coding for germ cell-specific proteins. The production of these primers and the resulting synthesis of the corresponding proteins characterises the dominant metabolic strategy of the germ cells.*

The presence of germinal granules and their accumulation in a specific cytoplasm is far from universal in the animal kingdom. In many groups it does not occur and the determination of the germ cells involves an inductive event [*Nieuwkoop and Sutasurya,* 1979, 1981]. In the axolotl, in particular, the primordial germ cells first appear in the lateral plate mesoderm, being induced from the ectodermal moiety of the blastula by the endodermal moiety along with the rest of the mesoderm [*Sutasurya and Nieuwkoop,* 1974]. Nevertheless, germinal granules have been observed in the equatorial region of the axolotl egg, i.e. in the prospective mesodermal area [*Williams and Smith,* 1971]. Although this is an isolated observation, it is a very important one: it suggests that the determination of the primordial germ cells may occur by two different mechanisms, which are also used in the determination of somatic cells. It also corroborates the idea that the germinal granules do contain stabilised specific primers. In fact, *the determination of the primordial*

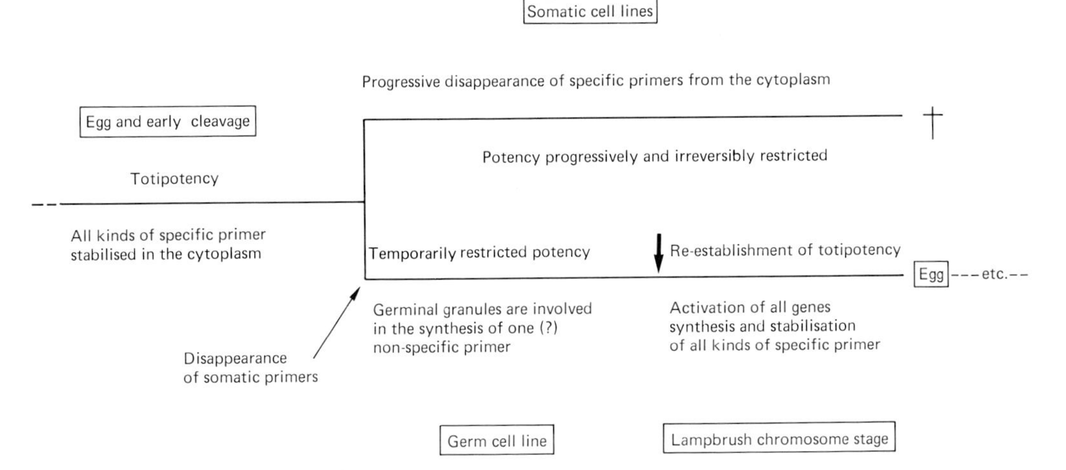

Fig. 22. Comparison of determination in the somatic and germ cell lines. Explanation in text.

germ cells would become effective when the concentration of germinal granules exceeds a certain threshold. When their concentration is not high enough from the start the blastomeres in question would engage in another differentiation pathway, unless particular extracellular information (an inductive event) enhances their production.

In the prospective area of a given germ layer the role of the determinant is restricted to the initiation of autonomous progression in a given direction (which sometimes ends in early cell death). The realisation of definitive differentiation always and everywhere requires in addition the summation of various sorts of extracellular information. Similarly, the segregation of the egg and sperm lineages, and later the differentiation of the gametes involving the completion of meiosis, require the registration of information of various kinds provided by the somatic cells. To be more precise, the presence of germinal granules is not the only factor needed to ensure the re-initiation of transcription during meiotic prophase. For instance, it is well known that in birds the primordial germ cells that fail to reach the gonad cytolyse. In the vertebrates generally the connection between the gonad and the mesonephros seems important: in isolated gonads meiosis does not take place [*Byskov*, 1974]. Also, the initiation of oocyte maturation requires the action of hormones, just as is the case for the completion of differentiation in many somatic tissues.

We thus note a striking parallel between the somatic and germinal cell lines as to the principles of differentiation (fig. 22): in both we find the triggering of autonomous progression by specific 'primers' produced in the oocyte, the progressive stimulation of a luxury metabolic strategy by the registration of extracellular information, and the repression of part of the genome. The essential feature of the germ cells would then be that they ultimately produce non-specific primers needed for the reactivation of all the repressed genes, something a somatic cell cannot do (with the possible exception of cancer cells which take on characteristics of embryonic cells). But to do this the germ line must be able specifically to reactivate the gene or genes which code for these non-specific primers. It needs its own specific primer or primers for this, which are probably typically contained in the germinal granules. It is obviously important that the presence of this primer or primers is not interrupted from the egg of one generation to that of the next, particularly during the period when the germ cells are restricted in their potency (fig. 22).

The Egg Antecedent Prepattern

The establishment of the EAPP consists in the successive determination of the a/p and d/v axes of the embryo (which do not always coincide with the polar axes of the oocyte). It results from the redistribution, in two steps, of a single type of cytoplasmic constituent, which controls the function of the GA clock (or may form part of it). The determination of the embryonic axes either results from the process of egg maturation or is due to external factors acting fortuitously.

Among the constituents that are heterogeneously distributed in the egg not all are involved in the establishment of the EAPP, but only those that take part in the determination of the first autonomous progressions. These are of two kinds. First there are constituents whose activity controls the functioning of the GA clock: the yolk granules and the mitochondria, which as a function of their concentration respectively influence the cleavage rate and the rate at which primers are made available (p. 111). Secondly, there are constituents which themselves form part of the GA clock: the RNA-protein complexes which contain the primers needed to re-initiate transcription, e.g. the granules present in the polar lobe of *Dentalium* (p. 105). Because several constituents of the egg (for instance various types of pigment granules) have no noticeable effect on the functioning of the GA clock, the features they determine in the egg when they are redistributed by ooplasmic segregation do not form part of the EAPP. This is demonstrated unambiguously by the development of the mammals. In the so-called 'dorsal' part of the uncleaved egg a strong alkaline phosphatase activity can be demonstrated, which later becomes strictly localised to the inner cell mass [*Mulnard*, 1955]. But we know that this group is exceptional in that the future fate of the blastomeres that deviate from the ectodermal pathway is determined exclusively by extra-embryonic information. Obviously, here the segregation of particular ooplasms plays no role whatever in the organisation of the blastocyst. Similarly, in certain regulative eggs particular ooplasms occupy the prospective area of structures to be induced later by an influence coming from the vegetal pole region, but because their constituents have no specific influence on the GA clock they can be eliminated without resulting in developmental anomalies.

The attention of embryologists has of course been attracted by those ooplasms which were the easiest to observe, and which consequently have become 'models' for the analysis of ooplasmic segregation. The formation of some of these may seem to be involved in the establishment of the EAPP while in reality it is only an accompanying phenomenon. For instance, the grey or yellow crescent of amphibian eggs does not form part of the EAPP because its removal along with the entire equatorial zone of the blastula does not prevent the development of a normal embryo [*Nieuwkoop*, 1969a]. However, its formation results from the same mass displacement of yolk platelets that is responsible for the symmetrisation of the 'endodermal moiety' [*Gerhart* et al., 1981]. In the eggs of the sea urchin *Paracentrotus lividus* an orange-coloured ooplasm occupies the area of the future endo-mesoderm [*Hörstadius*, 1939]. After treatment with lithium ions, which results in the over-production of endoderm, an enlargement of this ooplasm is observed. However, the blastomeres which contain it can be removed without leading to abnormal development [*Gustafson*, 1965]. Such seemingly contradictory results have made it difficult to reconcile the experimental data from different animal groups and to arrive at a coherent concept.

As we have shown, the entry of the early blastomeres into different autonomous progressions is usually effected by the desynchronisation of the activities of the animal and vegetal portions of the embryo (or, in exceptional cases, the accumulation of specific determinants towards the vegetal pole). Depending on the species this polarity may result either from the establishment of a metabolic gradient (as in sea urchins) or from the localised accumulation of a certain class of cytoplasmic constituents (e.g. the coarse yolk near the vegetal pole in amphibians). However, the initial organisation must be somewhat more complex to account for the later structure of the animal. For instance, if the 'endodermal' and 'ectodermal' moieties of the amphibian blastula were both entirely homogeneous the embryo would become radially symmetrical. That the chordomesoderm appears on one side of the 'ectodermal' moiety is due to the fact that the 'endodermal' moiety has previously acquired a dorso-ventral polarity [*Nieuwkoop,* 1969b]. This demonstrates a point of general validity. For all the major animal groups it is now assumed that *the structuring of the embryo has its starting point in the determination of two axes of polarity running roughly at right angles to each other* [*Nieuwkoop,* 1977, for the amphibians; *Nüsslein-Volhard,* 1979, for the insects; *Guerrier,* 1970a–c, for the molluscs and the annelids].

In some groups the same cytoplasmic constituents are involved in the establishment of both axes, and there is nothing yet to suggest that we have not to do with a general principle. In sea urchins [*Gustafson,* 1965] the properties particular to the animal region (high respiratory activity) extend further vegetally on the future ventral side. In the amphibians a simple rotation of the egg leading to a displacement of the yolk is sufficient to modify both the position of the 'endodermal moiety' and that of the chordomesoderm [*Pasteels,* 1964]. In ascidians, where the distribution of the mitochondria is a determining factor, they are less numerous in the vegetal region (the future endoderm) than animally, and less numerous ventrally than dorsally (where they are responsible for the determination of the muscle lineage). These features of the EAPP will later interfere with all the other organisational features that will appear during later readjustments and will thus confer on all parts of the organism (as well as on the organism as a whole) their antero-posterior and dorso-ventral polarities.

Usually when it leaves the ovary the oocyte already possesses an animal-vegetal polarity and often even a bilateral symmetry. This particularly holds for the amphibian oocyte, where the coarse yolk early on accumulates towards the vegetal pole, ascending higher on one side [*Wittek,* 1952]. In certain cases the oocyte acquires its polarity under the influence of the somatic cells by which it was surrounded. In the insects, for instance, the long axis of the egg and later that of the embryo coincides with the antero-posterior axis of the maternal organism. In sea urchins the point where the oocyte is attached to the ovarian wall becomes the vegetal pole [*Lindahl*, 1932] and later the posterior end of the larva. In gastropods of the genus *Lymnaea* six subcortical cytoplasmic localisations appear approximately in the equatorial region of the egg. They do not seem to belong to the EAPP but they define a symmetry plane which will also be that of the embryo [*Ubbels* et al., 1969]. Such observations have given rise to the idea that certain features of the organisation of the organism are already preformed in the oocyte, being 'imprinted' on its periphery (the 'cortex') by the somatic cells of the gonad [*Raven*, 1964]. However, the cytoplasmic constituents whose distribution defines one or two polar axes have not yet found their definitive places in the immature oocyte. They will unavoidably be displaced by the cytoplasmic reshuffling resulting from maturation, sperm penetration, amphimixis and the cleavage divisions. They can also be displaced by various factors of external origin, either under natural conditions or under the influence of various experimental manipulations. It follows from this that *the embryonic axes do not necessarily coincide with those of the oocyte and can often be determined at will by the experimenter.* The eggs of oviparous vertebrates provide a particularly clear example. The yolk platelets are heavier than the cytoplasm. Upon activation of the egg they become mobile with respect to one another and consequently tend to sink down. In amphibians a simple 180° rotation of the egg thus leads to inversion of the antero-posterior axis of the embryo because the yolk platelets accumulate in the former animal region [*Pasteels*, 1938]. In birds the orientation of the antero-posterior axis with respect to the long axis of the shell depends on the angle the egg takes up in the uterus [*Clavert*, 1962] or in the case of an egg taken from the uterus just after it has arrived there, on the position in which it is subsequently kept [*Kochav and Eyal-Giladi*, 1971]. Therefore, *even where the embryonic axes coincide with those of the oocyte, their determination is an epigenetic event.* The two main axes of the egg are not determined at the same time, nor are they determined in the same manner or by the same factor, even though their determination involves the redistribution of the same type of cytoplasmic constituent.

Both before and after fertilisation the activity of the mitotic apparatus results in displacements of cytoplasmic inclusions which contribute to the polarisation of the egg. Thus, it sometimes determines by itself the a/p and d/v axes of the embryo (again in two steps). During the maturation divisions the spindle takes up an eccentric position perpendicular to the oocyte surface: the animal-vegetal polarity is already established. Particular ooplasms may appear in connection with the deployment of the aster rays. For instance, in the *Lymnaea* egg the rays transport mitochondria and concentrate them near the animal pole [*Raven,* 1964]. Thus, in the molluscs the antero-posterior axis is established during the first maturation division. Nevertheless, its position may be modified at will by compressing the oocyte so that the spindle turns [*Guerrier,* 1970a]. In various species the development of the spermaster from the sperm centriole 'reshuffles' the cytoplasm in a direction perpendicular to the animal-vegetal axis. A similar movement occurs during the first cleavage division, when the spindle places itself in the equatorial plane (or in a latitudinal plane) in a position which depends on the localisation of the spermaster. In this way the egg acquires a bilateral symmetry determined by the sperm entrance point. This is sometimes retained by the embryo, as is the case in certain molluscs [*Guerrier,* 1970c]. Finally, even after the start of cleavage cytoplasmic movements may continue, as long as the blastomeres are still large. Sometimes the EAPP is established at this late stage; in a nemertean worm the constituents involved in the determination of the apical ciliary tuft and the prospective endodermal area are uniformly distributed in the uncleaved egg but are displaced towards the animal and vegetal poles, respectively, between the first and third cleavages [*Freeman,* 1976, 1978].

During maturation the superficial layer of the egg, called the 'cortex' (which comprises the plasma membrane and a subjacent layer of cytoplasm) acquires the capacity to 'attract' certain cytoplasmic inclusions or organelles. This is particularly clear in the insects, where this attraction is responsible for the formation of the syncytium, which will give rise to the blastoderm. The cleavage nuclei, each surrounded by a yolk-free island of cytoplasm and called the 'energids', move towards the superficial cytoplasm (periplasm) and fuse with it. This displacement, which also occurs in the absence of nuclei, fails to take place after lesion to the cortex [references in *Counce,* 1973]. More generally this property of the cortex, which it acquires at the end of maturation, manifests itself in the formation of more or less well-circumscribed ooplasms: the subcortical plasms. Certain kinds of inclusion such as granules, yolk platelets, etc. are concentrated there and are closely applied to the plasma membrane. For instance, in the eggs of *Aplysia,* a mollusc, granules of vitamin C, which are at first uniformly distributed, accumulate in a ring close to the animal pole at a specific stage of maturation, even if they were previously concentrated at the vegetal pole by centrifugation [*Peltrera,* 1940].

Cases are known where the appearance of subcortical plasms is clearly linked up with a pre-existing cortical organisation. In the egg of *Lymnaea,* for instance, the six subcortical ooplasms occurring approximately in the equatorial region are located where the nuclei of the follicle cells had been lying close to the cortex, and a polar ooplasm is located in the vegetal region that had been in contact with the ovarian wall. In certain sea urchins the cortical organisation can be directly observed: yellow pigment accumulates in a ring bordering the orange cytoplasm; in this area the lipid layers of the egg membrane are thicker [*Runnström*, 1928]. However, these are isolated examples which have nothing to do with the establishment of the EAPP. In fact, the cortex seems generally to adhere to certain constituents of the cytoplasm that happen to be or to come into its vicinity. In this way it may contribute to the stabilisation of the features of the EAPP. Experiments done on molluscan eggs provide beautiful examples. In *Dentalium* the positioning of the polar ooplasm seems to be determined by local properties of the cortex, for bacteria are always found attached to the egg membrane on the outer surface of the polar lobe [*Geilenkirchen* et al., 1971]. But other experiments performed later on eggs of various other molluscan species have shown that the cortex executes its attraction particularly on the granular determinants which come into contact with it. Certain species have a small polar lobe in which the determinants, in the form of granules, are uniformly distributed. The granules can be displaced by centrifugation, which has the same effect as polar lobe ablation. In other species with a large polar lobe no specific inclusions can be distinguished in the electron microscope and centrifugation has no effect. Deficiencies can only be obtained by treatment with cytochalasin B, which probably acts by detaching the determinants from the cell membrane [*Dohmen and Verdonk*, 1979].

This stabilising property of the cortex is also involved in the establishment of bilateral symmetry in amphibians [references in *Pasteels*, 1964; *Gerhart* et al., 1981]. During oocyte growth coarse yolk accumulates in the vegetal half of the egg; the peripheral yolk platelets adhere firmly to the cortex. Whenever a part of the cortex in contact with the coarse yolk is displaced in some way and comes to lie over the pigmented cytoplasm of the animal half, it carries along some of the yolk platelets, which thus form the so-called vitelline wall ('mur vitellin'). Seen from the outside this presents itself as a depigmented crescent, the grey or yellow crescent. This happens during normal development, when the entry of the sperm evokes the so-called 'cortical contraction' resulting in a movement of the cortex towards the animal pole, which is most pronounced on the future dorsal side, opposite the sperm entrance point [*Ancel and Vintemberger*, 1948; *Gerhart* et al., 1981]. But the same phenomenon occurs when the egg is rotated after fertilisation and the animal-vegetal axis kept at an angle for some time, because the heavy coarse yolk slides down along the cortex, leaving a 'tongue' of yolk platelets analogous to a vitelline wall; this side then becomes the dorsal side of the embryo, even if it was previously the ventral side. When a grey crescent had already formed before the experiment it does not disappear except upon UV irradiation [*Chung and Malacinsky*, 1980].

We have thus seen that factors of diverse origin can be involved in the establishment of one of the polar axes of the embryo. If all factors act together, only one of them ultimately determines the axis, overriding the others. This is shown particularly clearly by the observations and experiments on egg symmetrisation in amphibians summarised above. Conversely, under the influence of a single factor various different components of the cytoplasm reflect the establishment of the egg axis concerned, but not all of them are determinants for the corresponding embryonic axis. For instance, in the amphibians the penetration of the sperm has two visible effects on the organisation of the egg: (1) the appearance of the grey crescent, and (2) the formation of the 'dorsal yolk-free cytoplasm' in the dorso-animal quarter of the egg (probably in connection with the deployment of the spermaster) [*Herkovits and Ubbels*, 1979]. These two features are localised in the 'animal moiety' of the egg, but the dorso-ventral organisation of this 'moiety' is always established by the inductive influence emanating from the 'endodermal moiety' (the vegetal yolk mass). Thus, all the deployment of the spermaster does is to make more evident the dorso-ventral polarity determined by the entry of the sperm. Unlike in some oligolecithal egg types (p. 130) it plays no determinative role because here the activity of the GA clock is desynchronised more effectively by the displacement of the coarse yolk than by an accumulation of mitochondria. It is the entry of the sperm that symmetrises the vegetal yolk mass by evoking the 'cortical contraction', whose only visible effect, the formation of the grey crescent, plays no role in the programming of development.

Faced with the variety of phenomena that are involved in ooplasmic segregation one is led to doubt whether a general principle underlies the establishment of the EAPP. However, if one scrutinises the various factors that may determine polar axes in the egg one finds that they fall into two separate categories. In one group are the intrinsic factors, which originate from the properties of the oocyte itself – the activity of the mitotic apparatus and the physico-chemical properties of the cortex and the various cytoplasmic inclusions. The other group comprises extrinsic factors whose action is subject to chance: the point of entry of the sperm and the position of the egg in the gravitational field. In this view the programming of mammalian development – although it does not involve an EAPP but is based on an environmental prepattern – does not require the introduction of special principles. Just as in other groups both embryonic axes are determined by chance. The dorso-ventral axis is determined first by the appearance of the blastocyst cavity, which results from the differentiation of the trophectoderm cells that pump fluid from outside to inside [*Graham,* 1971]. The antero-posterior polarity is determined later by the angle the blastocyst makes with the uterine walls [*Smith,* 1980].

In conclusion, then, *the establishment of the EAPP does not require the presence of any pre-existing organisation in the oocyte. Neither does it seem to require any specific 'instructions', i.e. any transfer of special information from the somatic cells to the cells of the oocyte lineage.* The only instructions the oocyte receives are those that allow it to go through its maturation divisions, to accept the sperm, and to start its cleavage divisions. So far the establishment of the GA clock is not necessary, for we know that the activity of the nucleus is not required for fertilisation and early cleavage. To deliver their instructions the follicle cells must of course apply themselves closely to the cortex of the oocyte. They may leave an 'imprint' there by locally modifying its properties, which may in some cases contribute to the relocation of certain cytoplasmic constituents. Moreover, the oocyte must build up food reserves necessary for the survival of the embryo. If these accumulate at the pole that is closest to the ovarian blood vessels they push the nucleus into an eccentric position. When the instructions have been delivered the cytoplasm contains a molecular machinery which from the time of maturation onwards (and making a pause that ends with activation) can function without the registration of any extracellular information. The sperm does nothing but to restart this machinery; it does not contribute to it any specific molecular component, for its action can be mimicked by a simple prick of a needle. The functioning of the machinery inevitably results in internal cytoplasmic movements, which manifest themselves in the appearance of an axis of polarity that is sometimes determinative in the establishment of the EAPP: these movements are the reshufflings evoked by the development of the aster fibres during maturation divisions, amphimixis and cleavage, displacements bound up with changes in cortical properties as a result of maturation and sperm entry, and finally displacements due to the action of gravity on the yolk granules when they become mobile within the cytoplasm after fertilisation.

The Modalities of Cleavage

The physico-chemical and topographical transformations taking place in the cytoplasm of the egg after fertilisation have repercussions for the activities of the mitotic apparatus in successive divisions. Therefore, in those areas where precision is required for the 'interception' of a particular programme, the duration of interphases and the orientation of cleavage planes are necessarily adjusted to the progression of the GA clock and the establishment of the EAPP.

During the first cleavage stages the two asters of each mitotic apparatus are often of equal size so that the cleavage plane appearing at the equator of the spindle results in two equal-sized blastomeres. Moreover, as a result of the way in which the centriole replications succeed each other each cleavage plane is perpendicular to the two preceding ones (rule of 90° rotation). Finally, because interphases are absent the cleavage divisions succeed each other at regular intervals and are synchronous throughout the embryo. Some data are now available on the regulation of these events. Thanks to the technique of time-lapse cinematography it has been shown that each cleavage is preceded by a change in the consistency of the egg and more particularly by the propagation of a contraction wave through the cortex [*Hara* et al., 1980]. These events are also observed in fertilised or artificially activated eggs whose cleavage divisions are blocked by an antimitotic drug [*Hara* et al., 1980] as well as in enucleated egg fragments [*Sawai,* 1979]. In both types of experiment the events are virtually synchronous with those in normally cleaving control eggs. This clearly shows that the succession of cleavage divisions cannot be attributed to DNA replication in the nucleus, nor to the properties of the mitotic apparatus. It is due to an oscillatory event in the cytoplasm or the cortex, or both, which in this case is set off by activation but probably occurs in all cells; it has been given the name of 'cytoplasmic clock' [*Hara* et al., 1980].

After the usually equal, radial and synchronous initial phase the cleavage pattern soon becomes more complicated. There are three kinds of events. First, when the two asters of a spindle form in two cytoplasmic areas that are different, one is less developed than the other: the isolated mitotic apparatus has a truncated appearance [*Dan and Nakajima,* 1956; *Dan,* 1978].

As a result the spindle is asymmetrical and the cleavage plane separates two unequal blastomeres. Secondly, the cleavage pattern ceases to be radial when the spindles become tilted (no doubt as a result of changes in the cytoplasm [*Guerrier*, 1970a, b, c]) so that the new cleavage planes orient themselves at an angle to the preceding ones. Thirdly, the divisions become asynchronous. The cell cycle length of each blastomere is entirely determined by its own cytoplasmic machinery and not by some global signal: isolated amphibian blastomeres cleave in the same rhythm as in the intact embryo [*Hara*, 1977]. The role of the cytoplasm has been demonstrated directly. When in amphibians a portion of the egg cytoplasm is replaced by cytoplasm of another species, the egg cleaves in a rhythm that approaches that of the donor species [*Aimar* et al., 1981]. In mosaic eggs, when after the first cleavage the AB blastomere is made to incorporate a little of the cytoplasm of the CD blastomere, it cleaves in perfect synchrony with the latter [*Devriès*, 1973a].

All these data suggest that *the cleavage modes are influenced by the physico-chemical changes taking place in the cytoplasm as a result of activation and ooplasmic segregation.* Remarkable observations made on the eggs of the annelid *Chaetopterus* [*Pasteels*, 1934] demonstrate this directly. If the division of the nucleus is blocked it goes through a series of 'monaster' cycles in perfect synchrony with the normal cleavages. In addition the polar lobe appears during the first of these cycles, is resorbed and reappears during the second cycle, after which a sort of budding process mimicks the appearance of the first micromere quartet. Various experiments have demonstrated the importance of changes in cytoplasmic composition. When cells of the early sea urchin blastula are dissociated and cultured under conditions which preclude further differentiation, they continue to divide and to synthesise DNA in the same rhythm as at the time of dissociation, as if the system were 'frozen' [*de Petrocellis* et al., 1977]. In sea urchins [*Hörstadius*, 1935] and other invertebrates [*Freeman*, 1979] it has been shown that the position and orientation of the cleavage spindles depends on the time elapsed since fertilisation (i.e. on the state of the cytoplasm), and not on the number of cell cycles having taken place (i.e. on possible spontaneous, intrinsic changes in the 'cytoplasmic clock'). When in sea urchins the mitotic divisions are retarded by X-ray treatment [*Rustad*, 1960] or when one mitotic cycle is blocked [*Dan*, 1972], the micromeres appear at approximately the normal time. The influence of ooplasmic segregation on cleavage characteristics is already longer known [*Collier*, 1965]. When the ascidian egg is separated into two by centrifugation, one part contains all cytoplasmic inclusions, being deficient only in ground cytoplasm. This part cleaves normally [*Reverberi and La Spina*, 1959], whereas the other fragment fails to cleave. In amphibians inversion of the animal-vegetal polarity by 180° rotation of the egg also inverts the cleavage pattern: the blastocoelic cavity forms close to the former vegetal pole and the largest blastomeres are now those in the former animal area, where the coarse yolk has accumulated [*Stannisstreet* et al., 1980]. When the animal-vegetal axis is held horizontally the first cleavage spindle is oriented parallel to the boundary plane of the vegetal yolk mass, as in normal eggs, so that the cleavage plane bisects the yolk mass in normal fashion [*Hara*, personal communication].

Because they are determined by events which do not all influence the functioning of the GA clock, not all cleavage characteristics are necessary for development to proceed normally, but only those which are required to shunt certain blastomeres into a specific differentiation programme. The sea urchin egg, for instance, normally cleaves according to a very precise schedule. After the 3rd cleavage there are eight blastomeres of equal size, but at the 4th cleavage the spindles of the vegetal half move away from the cell equator and tilt, as a result of which each vegetal blastomere gives rise to a macromere and a micromere. The mitotic desynchronisation which begins at this stage is clearly linked up with the metabolic gradient already present in the unfertilised egg. The nearer one gets to the animal pole, the later the first interphases appear [*Parisi* et al., 1978]. However, the formation of the micromeres and the resumption of transcription in their nuclei are the only events which must be effected with precision to ensure that the cells are shunted into the 'endodermal' programme during the brief period it is available. Because the other blastomeres will 'catch' the 'ectodermal' programme anyhow, the orientation of their cleavage planes and their cleavage schedule is of little importance. When a treatment that reversibly blocks cell division is applied after the 2nd cleavage, the 3rd cleavage (which is normally equatorial) occurs at the time when control eggs perform the 4th cleavage (when the micromeres are normally formed). In that case the 3rd cleavage plane is displaced towards the vegetal pole but development is entirely normal [*Dan*, 1972].

We thus begin to see why the modes of cleavage are so well adjusted to the production of the 'ectodermal' and 'endodermal' programmes in the embryo. At any given time after fertilisation the activity of a given part of the egg is governed by two different kinds of organisation: one in the temporal dimension (functioning of the GA clock and various other physicochemical changes) and the other in the three dimensions of space (ooplasmic segregation, comprising the establishment of the EAPP). The activity that is specific for a given part of the embryo not only manifests itself in the nature of the programme that is available but is also reflected in the functioning of systems that are common to all cells: the mitotic apparatus, of which it determines the spindle orientation and the relative sizes of the asters, and the cytoplasmic clock, of which it modifies the cycle length. Consequently *the egg does not need a specific programme to regulate its mode of cleavage. When the oocyte has registered all the kinds of information it requires to accomplish its specific functions (production of primers, maturation, acceptance of the sperm and re-establishment of DNA transcription) the mode of cleavage is established ipso facto, just as are the modalities of ooplasmic segregation (which includes the establishment of the EAPP).*

Towards an Estimate of the Amount of Information Required for the Establishment of the Egg Developmental Programme

The amount of information required to produce a ripe ovum seems to be of the same order of magnitude as that needed for the establishment of a luxury metabolic strategy in a somatic tissue.

The comparison of the developing animal with a working automaton has given rise in the literature to a question that is rather unexpected for the embryologist: How large is the amount of information required for the establishment of the programme contained in the egg? It is obvious that an estimate, however precise, can contribute nothing to our understanding of how development is regulated; nevertheless, it is philosophically of some importance. All biologists who have compared animal organs to machines or man-made tools have been struck by the ingenuity of their design. It is an old ideal of cyberneticists to construct automata which can perform the most complicated kinds of work starting from the simplest possible single operation. The animal could be an example of such an automaton if it would have resolved the problem of the automatism of its development in as elegant a fashion as the problem of adapting its organs to their function. However, the few estimates that have been published give no ground to assume that this is so. *Raven* [1961] has suggested an amount of information equivalent to that contained in 6,000 books of 500 pages each; one would probably need fewer books than that even for the most minute description of the human organism at the morphological, cellular and molecular levels. The results to be discussed in this chapter will enable us to see that these orders of magnitude are far higher than corresponds to reality, because the estimates are not based on the correct theoretical considerations.

One of the most important conclusions of this chapter is that the oocyte is a differentiated cell just as any somatic cell. Because the egg developmental programme is a product of the oocyte's specific activity, it is established on the same principles as any tissue function whatever. Consequently, the comparison of the functioning of the somatic cell with that of a computer, which we proposed earlier (p. 24), must necessarily apply to oogenesis as well (fig. 23). The 'hardware' encompasses the 'arithmetical circuits' (the DNA) as well as an automatic machine to which the 'hardware' owes its capacity to reproduce itself: this consists of the centriole and the 'cytoplasmic clock', which regulates the frequency of the centriolar replications by rhythmic impulses. The clock's frequency can be modulated by the content of the cytoplasmic memory.

The 'software' consists of two components: (1) the 'input', i.e. the various kinds of extracellular information registered by the female germ cell lineage up to the oocyte stage, and (2) the content of the cytoplasmic memory, consisting essentially of the results of the 'processing' of the 'input'. Part of the content of the memory will be used to establish first the GA clock and then the EAPP (fig. 23, p. 140). It is evident, therefore, that *only the various kinds of extracellular information registered by the oocyte lineage, from the primordial germ cells to the mature oocyte, should be taken into account in evaluating the amount of information required for the programming of development.* If so far rather exorbitant figures have been suggested for this amount of information, this is because the developing animal has been compared not to a computer but to a decoding machine. Starting from the premise that the programme is encoded in the DNA one has quantified the amount of genetic information [*Apter and Wolpert,* 1965]. When the roles of the cytoplasm and the cortex of the egg were recognised one simply made an inventory of their constituent molecules and added the amount of information they represent to the amount contained in the DNA sequences [*Raven,* 1961]. One would make a similar mistake if, to evaluate the amount of information contained in the programme for a calculation, one would also take into account the information required to construct the computer as well as all the information stored in its memory.

It will be clear that only the 'input' needs to be taken into account for our purpose. However, in reality a complication arises if one would feel obliged to equal with 'input' all phenomena which experimental evidence suggests to be involved in the programming of development. What value should one accord to all those factors which, if absent, can be replaced by other factors (e.g. the role of sperm entry in the symmetrisation of the amphibian egg) and which usually act by chance? Rather than considering such stimuli received by the oocyte lineage and the egg, it seems preferable to consider the question from another angle: that of the problem the maternal organism has to solve to produce oocytes and to bring them to maturation;

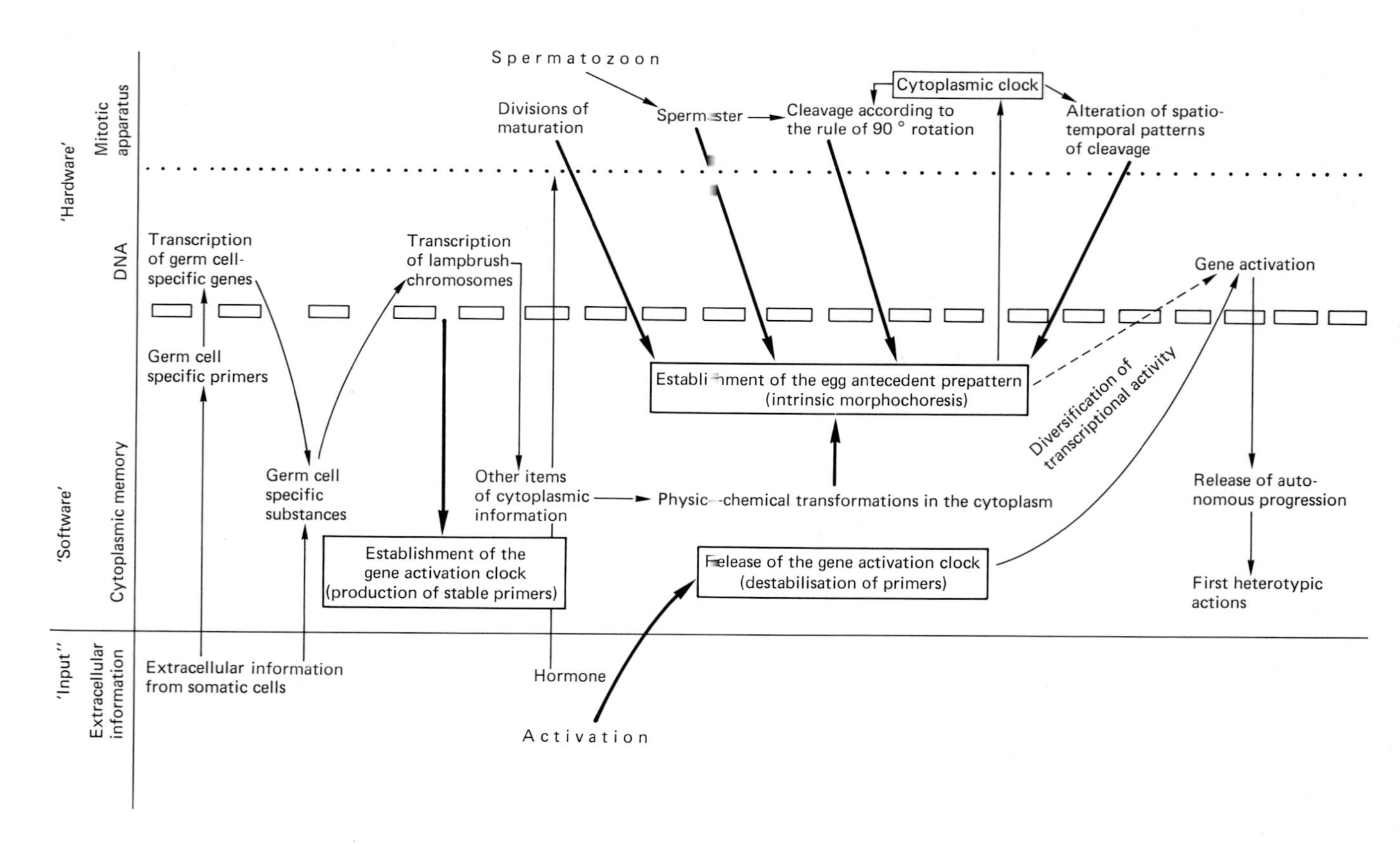
'Hardware'
Mitotic apparatus
DNA
'Software'
Cytoplasmic memory
'Input''
Extracellular information
Spermatozoon
Divisions of maturation
Sperm aster
Cleavage according to the rule of 90° rotation
Cytoplasmic clock
Alteration of spatio-temporal patterns of cleavage
Transcription of germ cell-specific genes
Transcription of lampbrush chromosomes
Gene activation
Germ cell specific primers
Establishment of the egg antecedent prepattern (intrinsic morphochoresis)
Diversification of transcriptional activity
Germ cell specific substances
Other items of cytoplasmic information
Physic-chemical transformations in the cytoplasm
Release of auto-nomous progression
Establishment of the gene activation clock (production of stable primers)
Release of the gene activation clock (destabilisation of primers)
First heterotypic actions
Extracellular information from somatic cells
Hormone
Activation

having been extruded from the ovary they are then by themselves capable of further development. The solution to this problem is the production and subsequent maintenance of some special classes of cells: the somatic cells of the gonad and the glands that produce the sexual hormones. By way of their specific biosynthetic activities these cell classes provide various kinds of extracellular information to the cells of the oocyte lineage in a certain temporal sequence. This is the only information that is specifically needed for this programming. Otherwise, the number of heterotypic actions (inductive events and effects of morphogenetic hormones) needed for the differentiation of the oocyte lineage is about the same as that involved in the differentiation of any other somatic tissue. This leads to the assumption that the amount of information required for the establishment of the programme of development is of the same order as that required for the establishment of any other dominant luxury metabolic strategy. It would be hazardous to attempt to quantify it, but judging from the simplicity of the principles on which the egg developmental programme is based it is probably not very large. If there really exists a single primer capable of reactivating the entire genome during meiotic prophase, its production would suffice to enable the oocyte to establish its GA clock. For the establishment of the EAPP it is not necessary that a detailed three-dimensional organisation be 'imprinted' on

---

Fig. 23. Diagram recapitulating the programming of development in the egg. The cytoplasmic memory of the primordial germ cell contains substances of maternal origin (among which are specific primers for the germ cells). Transcription as well as the registration of information deriving from somatic cells (particularly those of the gonad) contribute to changes in the content of the cytoplasmic memory, leading to the transcription of all structural genes on the lampbrush chromosomes. The results are (1) the storage of stable primers needed for the reactivation of genes during cleavage (establishment of the gene activation clock, the first component of the egg developmental programme) and (2) the storage of information needed for the initiation of maturation and the acceptance of the sperm. Maturation is the starting point for physico-chemical changes in the cytoplasm which manifest themselves particularly in the maturation divisions. The immediate effect of sperm entry is the re-initiation of mitotic activity: in the absence of specific cytoplasmic information centriole replications succeed each other automatically according to the rule of 90° rotation. The physico-chemical changes undergone by the cytoplasm, the deployment of astral rays during maturation and cleavage, and the impact of the sperm together determine cytoplasmic displacements (intrinsic morphochoresis) which establish the second component of the egg developmental programme: the egg antecedent prepattern. This component of the programme is responsible for desynchronisation in the sequence of destabilisation of primers and/or mitoses in different parts of the egg. As a result nuclear activities are diversified and the blastomeres are shunted into different autonomous progressions. This allows the first heterotypic actions to occur in the embryo.

the oocyte by its somatic environment. It arises spontaneously as soon as the physico-chemical conditions obtaining in the cytoplasm are made to change in a given direction.

The penetration of the sperm, which can be replaced by all sorts of parthenogenetic factors, does not seem to have to be added to the sum of the 'input' required for the programming of development. In the mature ovum the 'hardware' is so to speak 'set back to zero'. On the one hand all transcription on the DNA has ceased, although everything is ready for it to be resumed on any gene locus whatever. On the other hand, the mitotic apparatus of the oocyte has disappeared, and when the 'cytoplasmic clock' starts off again upon fertilisation its impulses will at first occur at regular time intervals, while the cleavage spindles will orient themselves according to the 'rule of 90° rotation'. The sperm completes the 'hardware' of the egg by adding to it its nucleus and centriole. By activating the egg it simply sets off a molecular machinery that is ready to function without requiring any further specific 'input'. The simplicity of the egg developmental programme and the probably small amount of information required for its establishment can be understood if we consider the manner in which cell activities diversify during development. We know that cells continuously communicate with each other, leading to an enrichment of their cytoplasmic memories. This can be clarified by means of a computer analogy (fig. 24). Imagine two computers designed in such a way that one starts working before the other, and that after each calculation each conveys part of its results to the other. If we assume for simplicity that their memories are empty at the start, we see that even with the same initial input A for both computers they will end up with different information contents in their memories (which amounts to structural amplification or cell diversification).

However, to complete its development on the basis of a simple programme the organism in reality uses another, complementary strategy that is peculiar to it. Three times during development the organism would attain an overall state of equilibrium, were it not that at that time a new cell or tissue property is introduced, enabling the registration of new information in the cytoplasmic memories of the cells, so that new readjustments can be initiated (fig. 25). During early cleavage the cytoplasmic information content already changes independent of any nuclear activity, a change terminating in ectodermal determination. The effects of activation would stop there were it not that at that time mitotic interphases appear. The resumption of transcription sets off the first autonomous progressions, which allow the first heterotypic actions to occur. As a result the cell surface properties change and diversify.

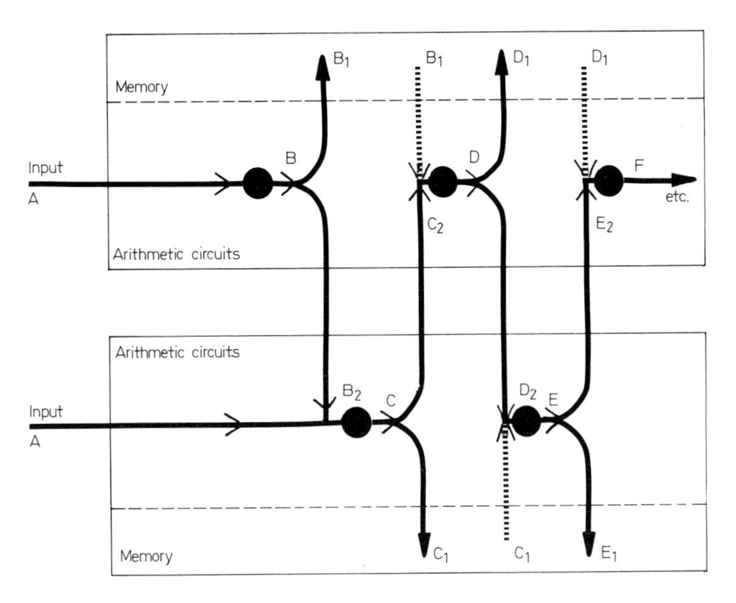

Fig. 24. Principles of structural amplification during development. Both computers begin to work one after the other with the same input A. In the first, part of the output B is stored in the memory (B_1); the remainder (B_2) is conveyed to the second computer and establishes the programme for its first calculations on the basis of information A. An output C is obtained, part of which (C_1) is stored in the memory, the rest (C_2) being conveyed to the first computer, and so on.

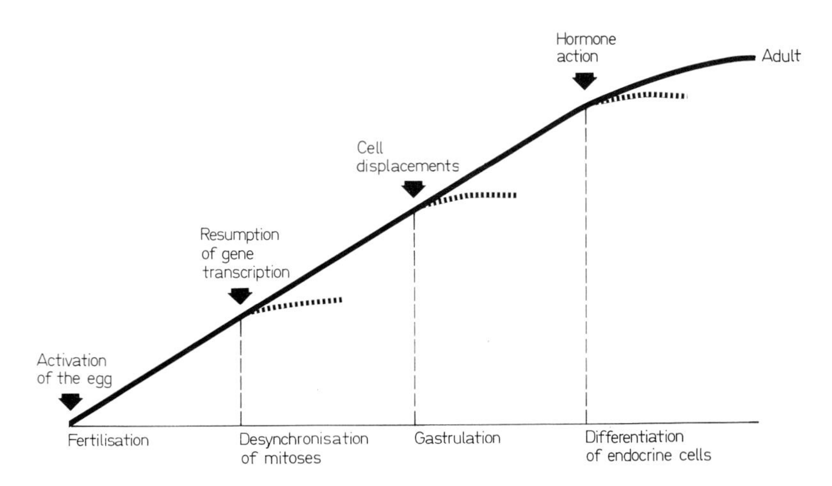

Fig. 25. The various changes in cell and tissue activity which successively restart the progression of differentiation (elevation of level of differentiation) during development. Explanation in text.

This results in cell displacements (morphogenetic movements), which bring together cell types that were previously widely separated. This allows inductive events to proceed. But organogenesis would also come to a halt before completion if it would not itself involve the differentiation of the endocrine glands and the vascular system. The circulation of the morphogenetic hormones is the factor that triggers the final readjustments of development.

Therefore, despite the many lacunae in our knowledge of the egg that remain to be filled, there is every reason to think that the programming of development involves less extensive devices than was thought until recently. Perhaps a few substances suffice to confer on the undifferentiated primordial germ cell its oocyte individuality. Although they no doubt determine specific cellular activities in the oocyte, they could be relatively simple in structure compared to the molecular complexes of differentiated tissue cells.

An erroneous concept of 'automation' too frequently leads to the idea that if there were a programme in the egg, the organism could only engender 'frozen' structures. This view was only recently maintained by *Gerhart* et al. [1982]. In our terminology development would then represent an inexorable concatenation of readjustments. However, with a small initial amount of information the organism can do much more than that. The constant exchange of information between the cells enriches the cells' memories and then prevents their contents from being erased. Consequently, when the relations between the cells are modified, this allows for 'improvisation' in the reproduction of structural patterns or for the remodelling of patterns already established. In many species the organism has at its disposal – on the basis of the same programme – a whole range of possibilities to adapt its form to the requirements of its mode of life. And what is more, the organism is capable of restoring its normal organisation after a part has been removed or displaced. This property, called 'regulation', has been used as an argument against the idea that development is programmed in the egg [*Brien*, 1974]. Because we take the opposite view – namely that the analysis of regulation is the best way to apply the principles propounded in this book – this will be the subject of our last chapter.

Structural Regulation as a Consequence of the Particulars of Developmental Programming

Judging by appearances only, the regulation of structure seems to be the spontaneous capacity for repair of organisms after they have suffered some alteration in their organisation. It seems to reflect a capacity of certain parts of the organism to reproduce the structural patterns of other parts. It is observed at all stages of the life cycle, from the egg to the adult. What is even more puzzling than regulation itself is the capricious manner in which it manifests itself. After elimination or displacement of a part of the organism, or after the addition of an extra part, sometimes the repair process fails to proceed to its end or results in an anomaly. Moreover, although there are species where all parts are interconvertible, both in the egg and the adult, usually this capacity is restricted to certain regions of the body, or to certain organs from a certain stage onwards. Generally, the capacity is ultimately lost at different times in the various parts and organs; usually the loss occurs earlier the more specialised is the adult. This rule is by no means absolute, however: certain species even show the inverse pattern.

Just as their emergence during development, the regulation of structures is the result of a 'group dynamics' of the cells; its mechanisms as well as its failures can only be explained on the basis of our knowledge of the properties of cell populations. We know that from a dynamic point of view embryonic and adult cell populations are quite different. The former are engaged in an autonomous progression which is accompanied by readjustments; the positional information registered by the cells results in progressive changes of the cell individualities. In contrast, in the adult the cells help each other to maintain their individualities by the exchange of information – which often ensures the maintenance of the tissues despite the possibility of dedifferentiation and despite cell renewal. This reminder will suffice to make clear that not all structural regulation is based on the same principles. We must distinguish two types, which must be analysed separately. The first is regulation of the embryonic type, which is observed in rudiments that are being formed, during both embryonic development and regeneration [*Chandebois,* 1973]. The second type is observed from the beginning of functional organogenesis.

Regulation of the Embryonic Type

In the embryo the restitution of deficiencies, intercalation upon association of originally different levels of the organism, the reduction of material excesses, and the assimilation of grafts are only possible if the operation has not led to alterations in the prepatterns other than topographical modifications that can be rapidly obliterated by the elementary social behaviour of the cells. The normal continuation of organogenesis after experimental interference does not constitute a repair process but reflects the execution of the developmental programme that has remained intact. The regulative capacity disappears as the structure of the organism becomes more complex.

Both in the egg and in the embryo the most characteristic manifestation of structural regulation is the normal continuation of development in spite of topographical modifications resulting from experimental interference. When a prospective area is removed its normal achievement is taken over by the remaining areas. This happens when a deficiency is created by removing or destroying part of the embryo (in which case some authors speak of 'restitution') or by recombining two parts that are normally spatially separated (in which case today the term 'intercalation' is used). When normal development follows upon the fusion of two identical rudiments one speaks of 'reduction of material excess'. When a part is removed and reimplanted in a different site it may develop in conformity with its new surroundings, in which case one speaks of 'assimilation'. Let us look at some of the most classical examples. In sea urchins either of the two first blastomeres, when isolated, is capable of forming a complete (though smaller) larva. A normal larva is also obtained when two eggs are fused crosswise after first cleavage or when the micromeres of the vegetal pole are recombined with animal blastomeres. Similarly, a normal limb may be obtained from a part of the prospective limb area or from two such areas fused into one. When a piece is cut out from the anterior part of the early neural plate, rotated through 180° and reimplanted, a normal neural axis will result.

When an operation has no influence on the normal course of development this must be because the readjustments involved have proceeded normally. Therefore, a priori there are two possible interpretations of what hap-

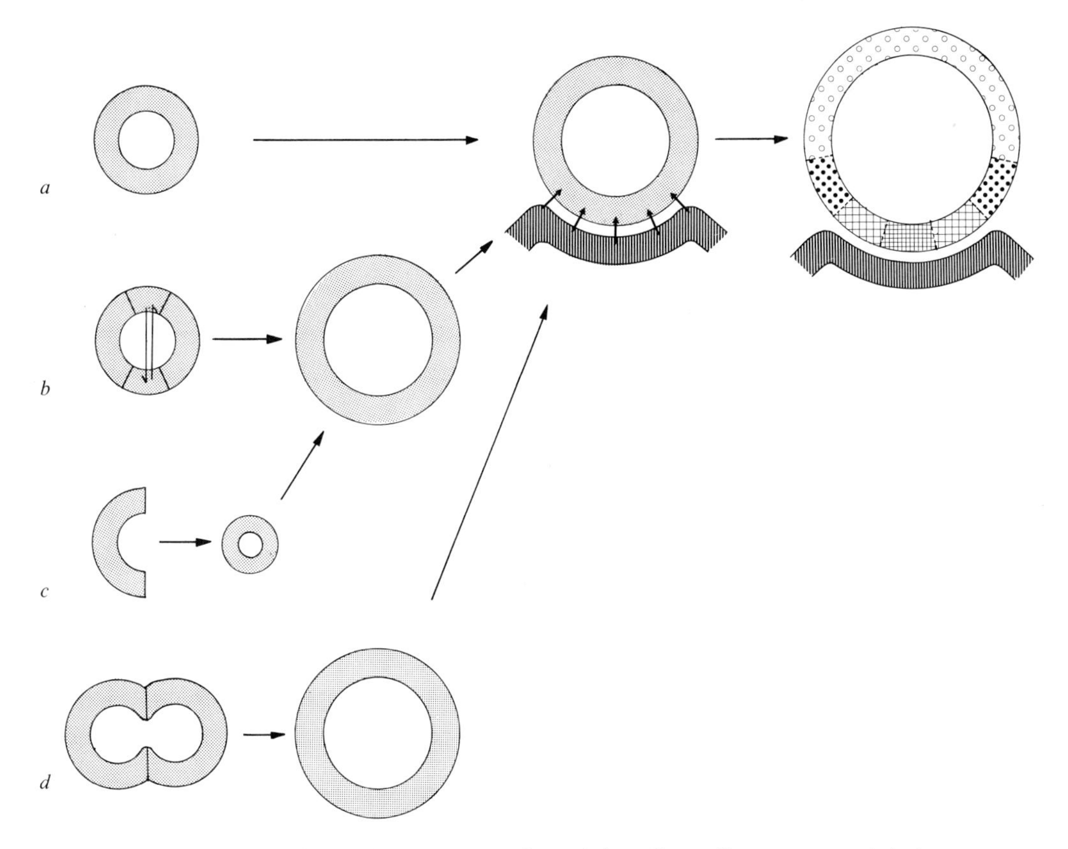

Fig. 26. Regulation in a homogeneous cell population. The readjustment proceeds in the same manner as in normal development *(a)* when the relative positions of constituent parts of the population have been changed *(b)*, when a deficiency has been created *(c)*, or when more of the same material has been added *(d)*.

pens in the cell population(s) concerned: either the programme of the first readjustment that follows upon the operation is not altered, or it is rapidly re-established.

We will start the analysis of the problem with the most simple situation: the operation disturbs a cell population of homogeneous appearance that is ready to perform a readjustment (fig. 26). If the population contains a covert pattern, we will assume that this will not manifest itself in visible structures, in other words, that the readjustment proceeds in the absence of an antecedent prepattern. If we remove a part of such a cell population and reimplant it in a different site or in the same site with altered orientation, this will not result in any change that can modify the programme of the next readjustment. For instance, we know that the neural plate is regionalised under the influence of the transforming action of the archenteron roof, whose covert graded pattern serves as imprinting prepattern, but that no pre-existing structure has been discerned in the ectoderm. Prior to the transforming action any part of the neurectoderm can be transferred to any other region without leading to anomalies (p. 74). Similarly, in the undifferentiated eye vesicle, which will later be organised under the influence of a positional prepattern provided by the epidermis and the mesenchyme (p. 36), an exchange of pieces between the future retina and pigment epithelium will leave the development of the eye unaltered.

Suppose now that we make a deficiency in, or graft extra material into, such a homogeneous cell population. Its size and shape are thus modified and if nothing would happen in time its next readjustment would take an abnormal course. The example of the eye vesicle will clarify the discussion. This will be transformed entirely into pigment epithelium unless it comes into contact with the epidermis, under the influence of which the cells retain the capacity to develop into retina. If a piece is removed and the eye vesicle fails to close, all cell types will appear but a normal eye will not be formed. If the vesicle closes without attaining its normal size, no contact with the epidermis will be established. However, we know that the eye vesicle and all other embryonic cell populations are perfectly capable of regulating their size and shape. We saw earlier that during an autonomous progression the mitotic activity of the cells is controlled by homotypic interactions: it is enhanced when the initial cell number is too small, and reduced when the number is too high, with the result that in both cases the final cell number is more or less the same. Moreover, if as a result of changes in cell surface properties the cells have arranged themselves in a certain manner, when the cell population is damaged or dispersed the cells have a tendency to rearrange themselves in the same manner. *Therefore, a homogeneous cell population, which is materially deficient as a result of ablation or in excess as a result of grafting, rapidly regains its normal state thanks to the elementary social behaviour of its cells. As a consequence, it is capable of re-establishing its intrinsic prepatterns before it loses its competence to interpret the extrinsic prepatterns.* Often this 'deadline' is delayed by the fact that cells isolated from their neighbours temporarily stop their autonomous progression, whereby time is gained for regulation.

A first example is furnished by the mammalian morula, which consists of identical blastomeres and gives rise to the blastocyst under the influence of an environmental prepattern (p. 106–107). Even if blastomeres are eliminated or several morulae are fused, a blastocyst of normal configuration can be formed, with the trophectoderm peripherally and the inner cell mass internally. Nevertheless, the number of blastomeres present at the time of determination of the blastocyst is of importance: the smaller it is, the smaller the chance that one cell is constrained by the others to take up an internal position and to form the inner cell mass [*Tarkowski and Wroblewska,* 1967]. In this case mitotic regulation has been demonstrated: a single embryo was obtained by fusing two blastocysts and it was found that the embryo was of normal size at the time the pre-amniotic cavity appeared [*MacLaren,* 1974]. A second example, that of 'successive polyembryony' in certain insects [*Vignau* et al., 1962], shows a case of regulation involving changes in cell shape. Normally the blastoderm forms the germ band in a region where there is a depression in the surface of the yolk beneath it, which allows the cells to form a prismatic epithelium. The rest of the blastoderm is a flat-celled epithelium and forms extra-embryonic ectoderm. When the germ band is removed the flat-celled epithelium stretches to re-establish blastodermal continuity, a property it shares with all epithelia [*Trinkaus,* 1967]. The cells that come to lie over the yolk depression change their shape and thus reconstitute the germ band. The operation can be repeated several times without leading to abnormal development.

Let us now consider the case where the cell population involved in the operation already possesses a visible or invisible structure that will be used as antecedent prepattern and as imprinting prepattern for neighbouring cell populations. *To ensure that the next readjustment is correctly programmed the structure in question should be rapidly reconstituted. Obviously this is theoretically impossible.* First, it is difficult to envisage how organisational features which have been progressively established in earlier phases of development due to the summation of prepattern effects can be instantly re-established. Secondly, such a reorganisation would always require a conversion of cell individualities, and we know that certain features of cell individuality, both visible and invisible, are irreversibly fixed, and with them certain visible or invisible features of organisation. In fact, it is possible in practice to follow the progression of determination in a rudiment by studying the loss of regulation capacity in it. For instance, the time at which the destinies of the future prosencephalon, rhombencephalon and spinal cord are definitively fixed is also the time when the various antero-posterior levels of the neural plate are no longer interconvertible. However, at that stage regulation is still possible within the prospective prosencephalon itself (p. 76).

All difficulties that have been encountered in the interpretation of experimental data are due to the fact that *not all invisible structural patterns are necessarily used for the establishment of a prepattern. In other words, not all invisible features of the cell individualities are necessarily used for the formation of visible structures during a readjustment.* Of those that are not so used the fact that they are irreversibly determined constitutes no obstacle for regulation. Nevertheless, there are cases where after an operation development is apparently normal but physiological anomalies appear later. For example, during the development of the prosencephalon there is a stage when antero-posterior inversion of polarity in the prospective area has no apparent effect but locomotion of the larva is disturbed [*Sládecèk,* 1960]. In addition, in those cases where an antecedent or imprinting prepattern is established as a covert pattern in the cell population concerned, not all features of this organisation are necessarily used. This holds particularly for graded patterns. For a normal programming of the readjustment it suffices that the two opposite ends are retained. *In this case the appearance of normal structures does not result from a reconstitution of the graded pattern but from interactions occurring during the readjustment,* which at first sight resemble intercalation (fig. 27, 28). A particularly illustrative example is the organogenesis of the nervous system, where the graded pattern of the archenteron roof serves as imprinting prepattern for the neuralised ectoderm during the 'transformation' phase (p. 75). An explant of competent ectoderm cultured with a combination of liver (which mainly induces prosencephalon and has weak 'transforming' capacity) and bone marrow (which has strong 'transforming' capacity and induces mesoderm) forms a complete neural axis [*Toivonen and Saxén,* 1955]. In sea urchins a metabolic gradient serves as egg antecedent prepattern. After experimental interference it is not only the embryos that contain all levels of the gradient which show normal development (such as those resulting from isolated meridional halves or from the fusion of two eggs), but also embryos

---

Fig. 27. Regulation in a cell population that organises itself on the basis of a covert graded pattern serving as antecedent prepattern. *a* Normal readjustment. One of the ends spontaneously acquires inductive properties (small crosses and upward arrows); the extent of the induced region (cross-hatched) is restricted by the action of the non-induced region (dots and downward arrows). *b* Transection parallel to the axis of the graded pattern. *c* Recombination of the two ends of the graded pattern. *d* Elimination of both ends of the graded pattern. In *b*, *c* and *d* the inducer can constitute itself (in *d* because there is still a sufficiently large difference between the two ends of the graded pattern). *e* Equatorial transection; the inducer does not form and consequently no further organisation takes place.

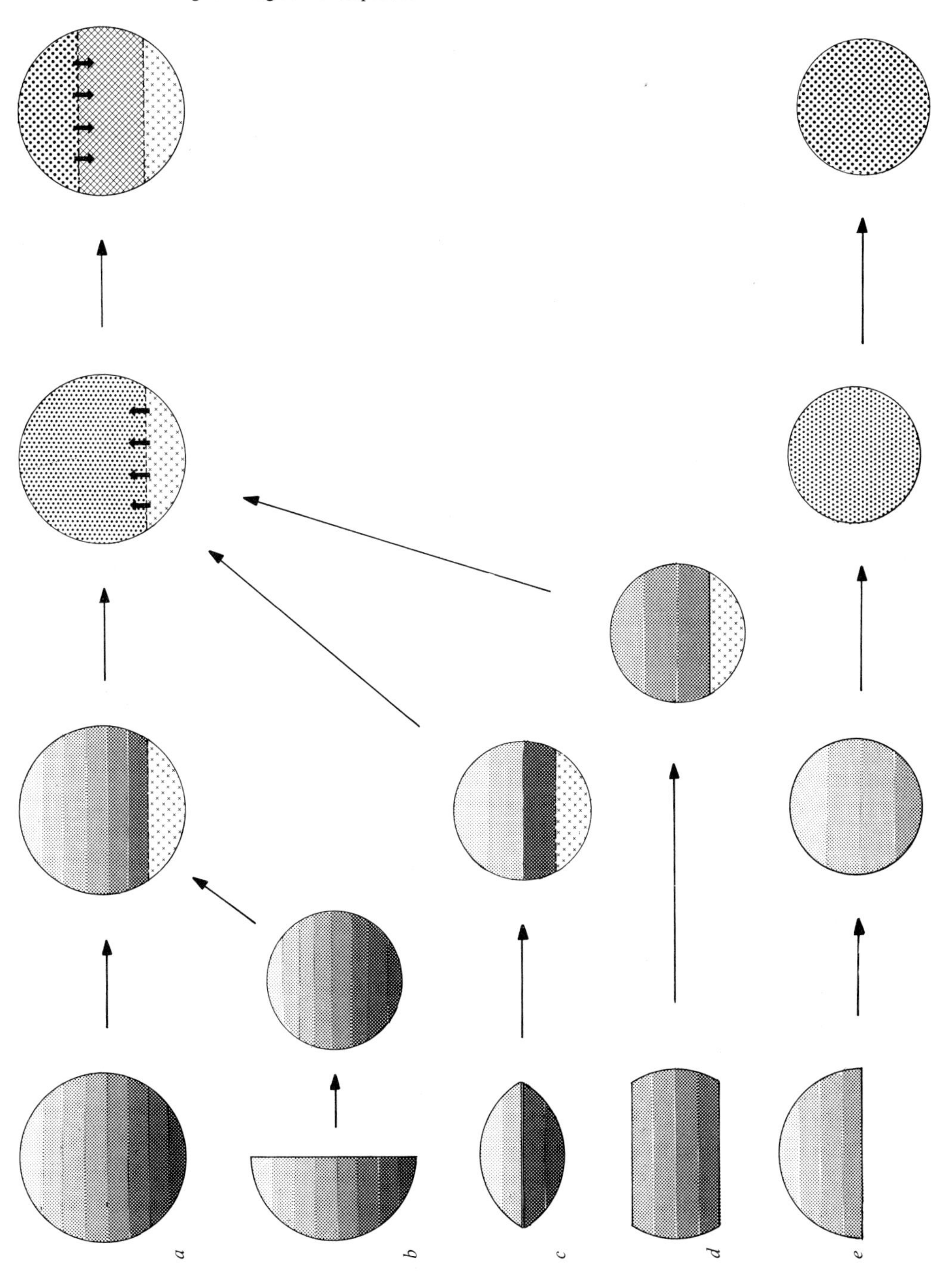
a
b
c
d
e

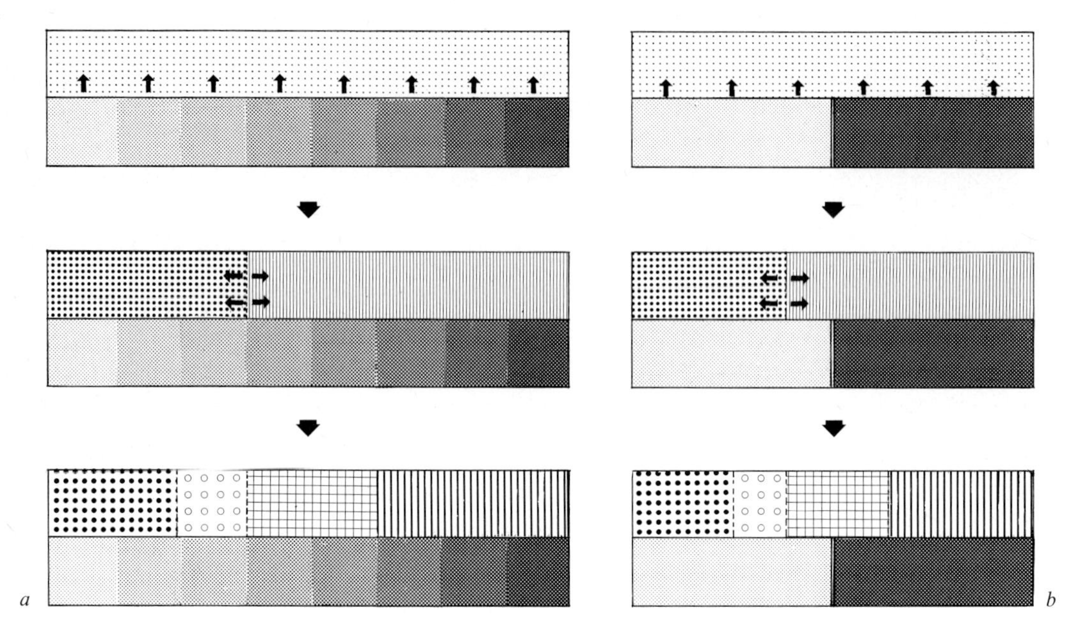

Fig. 28. Regulation in a cell population that organises itself under the influence of an inducer that serves as imprinting prepattern. *a* Normal development. The inducer contains a covert graded pattern. Two types of cells appear, which later interact to produce the definitive overt pattern. *b* The same readjustment is performed when the inducer is replaced by two tissues that have the same inductive properties as the two ends of the natural inducer.

in which there exists a large enough difference in metabolic activity between the two ends (for instance those resulting from the association of animal blastomeres and micromeres, or blastulae lacking only one of those two types of blastomeres). It is due to that difference that only one pole of the embryo acquires inductive properties, and it is probably the progressive nature of the induction and not the re-establishment of the gradient that is responsible for the appearance of the middle regions of the embryo. Nevertheless, the induction by itself cannot entirely explain regulation. The formation of normal structures despite the deficiency in inducible cell material suggests that an equilibrium is established between the positional information emanating from the latter and that coming from the inducer. This equilibrium, which is comparable to that in the normal embryo, then manifests itself in intercalation.

All these data lead to the idea that *not all features of an invisible organisation which constitute a given prepattern are of the same importance, because they do not all serve the same purpose.* Certain among them can be altered without precluding the normal course of a readjustment. *The most important features are those that foreshadow the most spectacular events: the emergence of specific organs. The more features in a prepattern play this role, the more aleatory is regulation.* This can be applied particularly to the case of the egg antecedent prepattern. Although this always comprises two perpendicular polar axes, not all regulative eggs show the same regulative responses to the same experimental interference. A comparison between sea urchins and amphibians is particularly instructive. In the former the animal-vegetal gradient foreshadows the determination of the three germ layers but the future dorsal and ventral sides do not differ as to the tissues they will form later. Whenever an operation leaves the difference between the animal and vegetal poles intact (which is particularly the case for isolated blastomeres or embryos fused after the first cleavage), a normal embryo is obtained. In amphibians this same condition must be fulfilled, but in addition the determination of the chordomesoderm, which will function as the organisation centre of the embryo, requires a sufficiently large difference between the future dorsal and ventral sides. When the first cleavage plane is 'sagittal' each of the two blastomeres comprises part of the dorsal and ventral sides of the egg, as determined by the symmetrisation of the vegetal yolk mass (p. 129). When they are isolated they both form a complete embryo. This is not so when the first cleavage plane is 'frontal', giving rise to a 'dorsal' and a 'ventral' blastomere. When the two blastomeres are separated during cleavage by a ligature, both of them can form either a whole embryo or a 'belly piece' which does not develop further [*Dollander,* 1950]. It seems that it is a matter of chance whether the ligature accentuates or obliterates the dorso-ventral polarity, presumably depending on how it displaces the coarse yolk relative to the cortex. In another experiment two 4-cell embryos were fused into one; sometimes a normal embryo formed but in other cases three blastopores appeared, apparently because the 'dorsal material' of the two embryos had become distributed into three separate regions [*Mangold and Seidel,* 1927].

Another difficulty in the analysis of regulatory phenomena in the embryo arises from the fact that *an operation may damage a cell population that is engaged in a readjustment.* Its organisation is already modified compared to that present at the time of determination, but it is not yet the definitive organisation characteristic of the end of the readjustment. The operation blocks the registration of part of the positional information by the cells, and the way in which development will continue depends on the information content of the cell memories at the time of operation: consequently, the development of the isolated part will suggest a reorganisation. This has often been observed in cell populations through which a progressive induction is propagating and which were separated from the inducer during the process. The cells which have not yet undergone all the successive determinations will then perform the normal achievements of cells which normally lie further away from the inducer, and this has sometimes been erroneously interpreted as a change in level specification, i.e. as regulation. For example, the prospective area of the kidney, isolated at the middle neurula stage, forms blood as well as kidney tubules [*Yamada,* 1940]. The most striking cases are those where a part of a cell population at the time of isolation temporarily shows an organisation that is very similar to that which the whole population would have shown at the end of the inductive event. Because this organisation serves as antecedent prepattern for the next readjustment, development may continue although it is never entirely normal. A case in point is the behaviour of the isolated middle zone of the axolotl blastula [*Nieuwkoop,* 1969 a]. It comprises part of the prospective chordomesoderm and endoderm, but because at the time of isolation it still partially consists of prospective ectoderm it produces a complete (though malformed) larva.

In certain experimental situations embryonic regulation may manifest itself in duplications. In the simplest case a rudiment or a region of the embryo is split into two parts which are permanently prevented from fusing. Each of the parts undergoes restitution as if the other part had been eliminated. An example is provided by experiments on the eye vesicle, which when split into two gives rise to two independent eye vesicles which each exert normal inductive actions on the surrounding tissues, in particular inducing two separate lenses [*Reinbold,* 1958]. The well-known *duplicitas anterior* embryos obtained by *Spemann* [1900] as a result of ligating the gastrula in the sagittal plane constitute a similar phenomenon. The ligature hinders the forward expansion of the archenteron, deviating the migrating cells on either side of it. This results in a bifid notochord rudiment and two separate archentera. Because the walls of the latter have the same conformation as a normal anterior end the inductive events succeed each other as in the normal embryo, hence the formation of two symmetrical heads. In other cases duplication results from intercalation: a graft of excess material is placed in a cell population competent to respond to its organising influence. It thus creates for itself an environment that is identical to that it would have had in the normal embryo. Sometimes this involves progressive induction. For instance, in amphibians the transplantation of a piece of chordomesoderm into the ventral mesoderm elicits the emergence of a (partial) supernumerary embryo

[*Spemann*, 1918]. But other kinds of heterotypic action can have similar effects: a duplicated limb can be obtained by grafting an apical ectodermal ridge to the base of an early chick limb bud, where it brings about localised outgrowth of the mesenchyme and ultimately the formation of precartilage condensations [*Zwilling*, 1956].

Suppose we have a number of computers which function jointly according to a common programme by communicating their 'output' to each other, and that an operator modifies the connections between some of them. For the final result to remain unchanged the operator would also have to erase the memories of the computers in question and to feed new 'input' into them. Obviously this operation would not have been programmed from the start: it would constitute a 'repair' prompted by an unforeseen event. In the embryo nothing can happen spontaneously that was not programmed in the egg, and consequently any 'repair' in the strict sense is impossible. *What the experimenter perceives as a return to normal, involving a sort of 'improvisation', is in fact the continued execution of the developmental programme, the course of which was not modified by the operation. The reason for the illusion is that development is itself based on continuous controls which ensure equilibrium between the cellular activities, in other words on regulation.*

The autonomous progression of differentiation in embryonic cell populations is a sort of self-regulation. Thanks to the homotypic interactions which regulate mitotic activity and specific cellular metabolism, each autonomous progression is completed on the basis of a certain number of cells and a specific cell arrangement. The properties which enable a cell population to acquire a specific size and shape in the normal embryo are the self-same properties which enable it to re-establish these following a deficiency or the addition of excess material. In other words, *any cell population engaged in autonomous progression is a self-regulating system. The establishment of new features of organisation as a result of the readjustment of cellular activities also implies self-regulation.* This is particularly evident in the case of progressive induction. The exchange of information between the cells that are transformed by the inducer and those that fall outside its range of influence (p. 152) leads to a state of equilibrium which ensures the emergence of typical structures, also when the number of cells involved is increased or decreased. However, once this self-regulation is completed development can no longer follow a normal course after experimental interference. Once abnormal features have been introduced in one region of the organism, they cannot but be amplified from one readjustment to the next and can never again be obliterated. The situation is comparable to that which results from a mutation or from the action of a teratogenic substance, except that here the experimenter changes the antecedent, imprinting and positional prepatterns and not the elementary social and environmental prepatterns (p. 89, 96).

As a result of the increasing parcellation taking place in the phase of organogenesis the normal continuation of development after operative interference becomes more and more aleatory. To avoid altering antecedent, positional or imprinting prepatterns the interference must be restricted to smaller and smaller areas. Finally, at different stages the various organ rudiments become altogether incapable of reconstitution, except when the appearance of new tissue properties opens up the possibility of regulation of a new type: adult regulation.

Regulation of the Adult Type

Regulation of the adult type, which essentially takes the form of regeneration by intercalation, involves the conversion of one part of the organism into another. The tissues of the part in question must possess the necessary potencies, and an adequate programme of cellular interactions must be furnished to the cells involved in the conversion. In so-called terminal regeneration it is the process of wound healing that establishes the programme, which is then executed by dedifferentiated cells.

In not a few animal forms individuals which are past the stage of prefunctional organogenesis – and indeed even adults – are capable of structural regulation which is at first sight reminiscent of embryonic regulation. When a deficiency is set normal organisation may be re-established; excess material may be resorbed; grafts or incisions may result in partial reduplication of the individual or of one of its organs. Particularly simple examples are furnished by the planarians, those flatworms which have been among the most frequently used materials for the study of adult morphogenesis. A complete worm can usually be obtained from a piece isolated from almost any level of the body, or by associating a head piece with a tail piece. Grafting an extra head in the posterior region almost always results in a partial reduplication of the animal, more particularly in the appearance of a second pharynx.

Despite the superficial analogies, however, adult regulation differs fundamentally from embryonic regulation in its visible manifestations [*Chandebois*, 1973]. When the organisation has returned to normal after a deficiency was created, or in the case of a reduplication of the individual, one always finds newly formed structures juxtaposed to pre-existing ones. Two modes of adult regulation are known (which may occur together in one organism): (1) morphallaxis, i.e. a partial remodelling of old structures, and (2) epimorphosis, i.e. the formation of structures de novo from a mass of undifferentiated cells called the blastema. In both cases, and contrary to embryonic regulation, *we have to do with a true repair process:* cells which were until then integrated in a certain part of the organism are converted to form another part. Therefore, the current use of a special terminology for adult regulation is wholly justified. One does not speak of restitution or intercalation but of 'terminal regeneration' and 'intercalary regeneration'.

Often organisms which are still developing regenerate better than adults. In rodents, for instance, complete reconstitution of the foot is possible in the fetus but becomes impossible after birth [*Aizupet*, 1935]. Taken together with the superficial analogies mentioned above, this observation would seem to support the idea that regulation capacity in the adult represents the persistence of embryonic regulation capacity, either in the whole organism or in certain parts only. However, it has been shown that sometimes a certain time elapses between the extinction of embryonic regulation capacity and the appearance of the faculty of regeneration. In amphibians there is a period during development when reconstitution of the tail bud is no longer possible but the faculty to regenerate a tail has not yet appeared [*Vogt*, 1931]. Similar observations have been made in planarians: head regeneration is only possible after the brain has differentiated, which is some time after the loss of embryonic restitution capacity and the first appearance of the nervous system [*Bardeen*, 1902]. *Obviously regeneration only becomes possible when the tissues have reached a certain degree of differentiation.* At first sight this would appear to be in formal contradiction to the fact that in the embryo it is the progressive structuring of the cell populations that brings with it the decline of regulative capacities. In actual fact, however, it is the completion of differentiation that ultimately leads to the reappearance of regulative capacities, simply because as a result of fundamental changes in tissue properties and the stabilisation of cell individualities a new possibility emerges: that of true *repair.*

Contrary to what has long been thought, organisms that regenerate well (with the exception of the hydroids) do not contain reserves of undifferentiated and totipotent embryonic cells which would be able to build up missing parts de novo [references in *Chandebois,* 1976a, c]. It is the differentiated tissues themselves that furnish the necessary material. The cell transformations involved are possible because in an intact tissue the maintenance of dominant luxury metabolic strategies and of other features of the cells' individualities as a rule requires constant exchange of positional information. As a result of a change in its environment a tissue can embark on the specific activities of another tissue, within the limits of the histogenetic potencies it has retained during development. Various mechanisms may be used for this, the one most frequently encountered being that of transdifferentiation, which requires a prior transitory dedifferentiation (p. 17). Another possibility is tissue modulation, which is effected through replacement of worn-out cells from a stock of generative cells (p. 48).

If we leave the hydroids and planarians out of consideration (where the differentiated cells remain totipotent), the possibilities of transdifferentiation and modulation become more or less restricted during development, to different degrees in different species and tissues. It follows from this that the formation of a regenerate usually requires the simultaneous involvement of several tissues of the stump, the activities of which must be rigorously coordinated. To explain the mechanism of the formation of a regenerate we have earlier introduced the concept of *cell transformation system* [*Chandebois,* 1965, 1973, 1976c].

A cell transformation system can be defined as an assembly of various tissue types – or various cell types belonging to the same tissue – which are interconvertible: this property is due to the presence in the system of a single class of undifferentiated cells sharing the same histogenetic potencies – these are the 'generative' or 'dedifferentiated' cells, so named according as they do or do not belong to a renewing cell population (fig. 29a). Generally the stability of visible structure reflects the existence of a numerical equilibrium between the various categories of cells. Within a cell transformation system this implies the regulation of cell division, the maintenance of those features of cell individuality that have remained reversible, and the control of differentiation during tissue renewal, all of which are ensured by the constant exchange of positional information.

Because undifferentiated cells in an adult organism are usually restricted in their histogenetic potencies, the organism has as many cell transformation systems as there are undifferentiated cell types. In each of these the numerical equilibrium is in part maintained thanks to information coming from neighbouring systems (fig. 29b). The amphibian limb furnishes a good example. It comprises the epidermis on the one hand, and various tissues of mesodermal

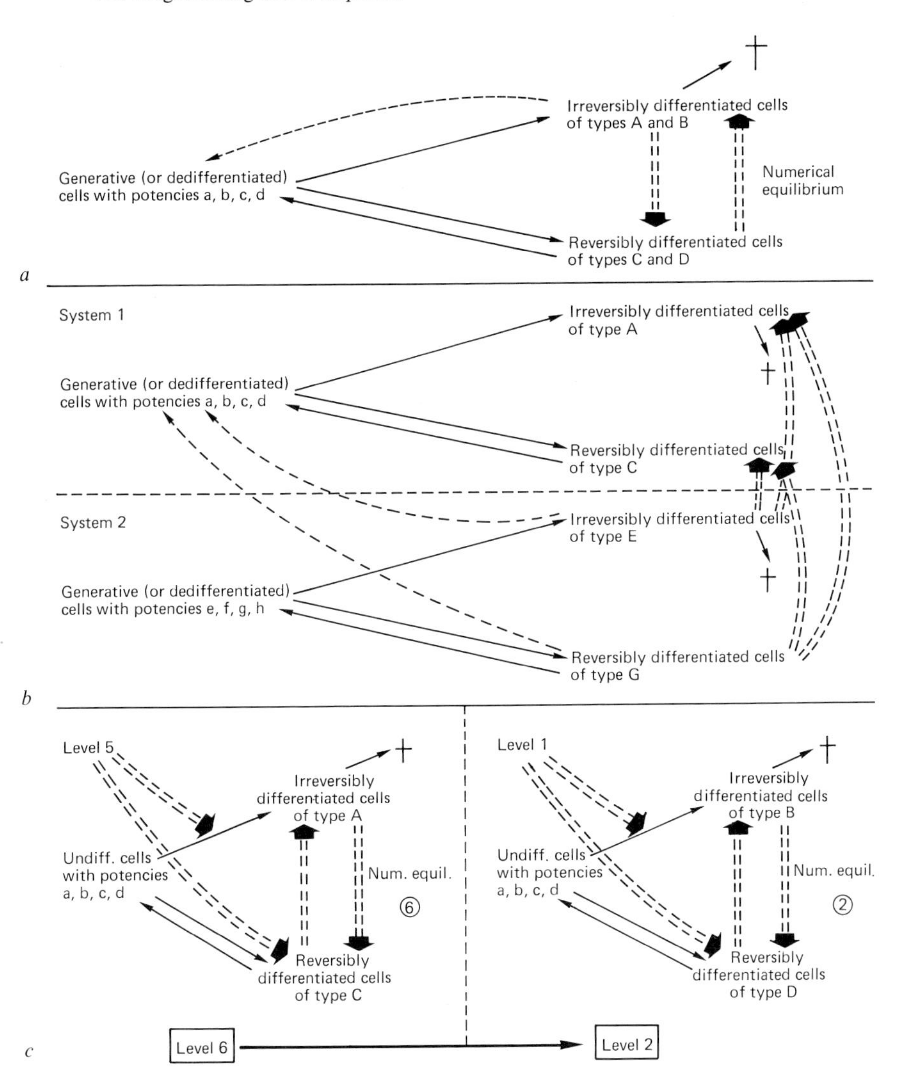

Fig. 29. Cell transformation systems. *a* Controls maintaining the characteristic cell type relationships within one system. *b* Controls of the cell relationships in system 1 by the differentiated cells of system 2. For simplicity the analogous controls exerted by system 1 over system 2 are not shown. *c* Transformation of level 6 into level 2 of the same system by a shift in the cellular equilibrium resulting from the replacement of the adjacent level 5 by level 1. ---▶ Control of mitotic activity; ===▶ control of specific syntheses.

origin on the other: connective tissue, muscle, cartilage and bone. The epidermis cannot arise from these tissues nor can it produce them: it constitutes a separate system and all differentiated cell types it contains originate from the same generative cells. All the mesodermal tissues together constitute the second system, for under various experimental conditions probably each of them, after having given rise to the same type of undifferentiated cell, can give rise to any of the others [*Trampusch and Harrebomée,* 1965; *Wallace,* 1981]. Nevertheless, the mesodermal tissues dedifferentiate when the epidermis is removed, and the epidermis shows abnormal proliferation when it is brought directly into contact with muscle tissue instead of dermis [*Seilern-Aspang and Kratochwil,* 1965].

The organisation of most cell transformation systems reflects variations in the cell equilibrium from one region to another. If one considers an arbitrarily delimited region within the system, its cells constantly register information coming from cells of the adjacent regions in the same system, or sometimes from those of a neighbouring system. Consequently the specific type of cellular equilibrium that characterises a given part of the organism necessarily depends on the specific equilibria of the adjacent parts. The continuous regulation of cell activities that ensures the stability of acquired form at the same time makes structural regulation possible. As a result of a change in its surroundings the equilibrium in a given region can be shifted to take on the characteristics of another region, without dedifferentiation being indispensable (fig. 29c). This is how morphallaxis comes about. For instance, when a planarian is transected behind the pharynx the posterior piece forms a blastema that only reconstitutes the head. The presence of the head evokes a remodelling of the stump tissues, as a result of which the missing regions reappear, particularly the pharynx. When a given region is suddenly deprived of the normally adjacent region a massive dedifferentiation of cells results, followed by 'cellular activation' (p. 14). The utilisation of dedifferentiated cells to form missing structures de novo represents the principle of epimorphosis. The dedifferentiated cells can form other structures, but only if they receive the necessary information. Lens regeneration in urodeles is a well-known example. As a result of removal of the original lens the iris cells dedifferentiate, but they do not form a lens under any environmental conditions whatever. For instance, they fail to do so when they are transplanted to the brain [*Yamada,* 1967]. Moreover, if transdifferentiation occurs outside the environment of the eye the result is a small lentoid much different from a functional lens and analogous to the lentoids formed in vitro [*Eguchi and Okada,* 1973].

If the part that must be replaced by epimorphosis or morphallaxis (or both) comprises several cell transformation systems, all of them must be present in the region affected by the operation, so that each will contribute its share of dedifferentiated cells to the regenerate. If one of them is absent the corresponding tissues will be missing in the regenerate. For instance, in annelids it is known that if the endodermal tissues are eliminated near the amputation plane the regenerate will fail to form a digestive tract [*Abeloos*, 1950]. In addition to this, however, because of the interdependence of the systems in their differentiation the absence of one of them usually prevents the other systems from organising themselves normally in the regenerate; this holds for instance for the central nerve cord in annelids [*Avel,* 1932; *Boilly and Combaz,* 1970].

In an organism or organ one can always arbitrarily define a number of levels arranged along one polar axis (e.g. antero-posterior or proximo-distal). Every regeneration process automatically produces a part of this sequence of levels (or sometimes the whole sequence). This occurs according to a 'repair programme' which allows for the conversion of the cellular material of the stump and determines its modes. To fix our thoughts, let us suppose that a transection is made between levels 5 and 6. For the cells of level 6 to reform the sequence 1–5, either by epimorphosis or by morphallaxis, only one solution is theoretically possible: they must be brought into contact with level 1. As a result they will receive the information that enables them to form level 2, which will enable other cells to produce level 3, etc. In this way a repair programme is established as soon as two levels that are normally widely separated in the system find themselves side by side: all intervening levels will be reconstituted. These theoretical predictions tally very well with what is known about 'intercalary regeneration': the case where the regeneration process is programmed by the experimenter himself by joining together two parts of the same system that are normally separated by other parts [*Okada and Sugino,* 1937, for planarians; *Bullière,* 1971, and *French* et al., 1976, for the insects]. A regenerate is *always* formed; and it always encompasses the whole sequence of levels that normally lies between the section surfaces confronted, irrespective of the orientation of the pieces (fig. 30). For instance, one has made the operation in such a way that a piece was joined with its posterior (or proximal) surface to the anterior (or distal) surface of another piece coming from a more anterior (or more distal) region of the same system. In this case the intercalary regenerate has its polarity reversed with respect to the polarities of both of the joined pieces. This polarity reversal does not only concern the position of morphological structures (in planari-

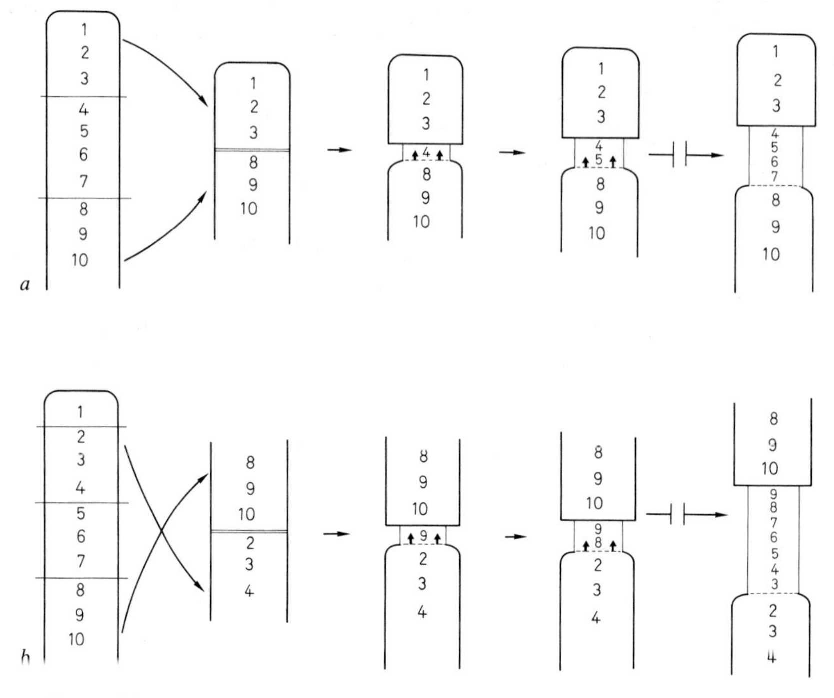

Fig. 30. Diagram showing the reconstitution of levels during the formation of an intercalary regenerate of normal *(a)* or reverse *(b)* polarity.

ans the pharynx is reversed with the mouth at the anterior end) but also the cell properties (in planarians the epidermal cilia beat in a direction opposite to normal). In contrast, when two pieces are joined with their homologous levels they both remain incomplete, irrespective of their orientation: no regenerate appears.

What happens if we remove one end from a system consisting of a number of 'levels' along a polar axis? It will be obvious that the cells that dedifferentiate near the amputation surface will be unable to form a regenerate unless some topographical alteration occurs that provides them with new positional information. If cellular displacements would occur at the amputation surface which result in features that are characteristic of the most distal missing level (which amounts to 'distalisation' [*Faber,* 1965]), this would *automatically* lead to the appearance of all intervening levels by intercalary regeneration. The only cell displacements that do occur are those involved in the closure of the wound. It has now been firmly established that they indeed constitute the sole factor involved in the establishment of the regeneration programme. In many different systems such as amphibian limbs [*Lheureux,* 1977], insect limbs [*Bart,* 1969; *French,* 1976], the parapods of marine annelids [*Boilly-Marer,* 1971], and planarians [*Schilt,* 1970; *Chandebois,* 1979], it has been shown that the emergence of the blastema is the result of the confrontation at the amputation surface of epidermis from two opposite body surfaces (e.g. dorsal and ventral epidermis): this amounts to the reappearance of the topographical characteristics of the most distal level to be regenerated, because it is only there that the two kinds of epidermis normally meet. The confrontation occurs spontaneously during wound healing after amputation but can also be brought about experimentally, in which case it leads to outgrowth of a supernumerary distal end without removal of the normal one [references in *Chandebois,* 1976 a, b]. The confrontation of dorsal and ventral epidermis is not sufficient, however. It must be followed by the formation of a wound epithelium stretched out between the dorsal and ventral borders of the wound [*Thornton,* 1968; *Chandebois,* 1979]. Wound healing is not only indispensable for the emergence of the blastema but it furnishes to the dedifferentiated cells information needed for the organisation of the regenerate. This has been shown by experiments on planarians [*Chandebois,* 1980 b], which have the advantage that they can produce three different types of regenerate, cephalic, caudal and lateral, at all levels of the body, depending on the direction of transection. The distal end of the regenerate represents the 'hinge' between the dorsal and ventral body surfaces. In the normal animal it is only on the lateral borders that this hinge coincides exactly with the boundary between the dorsal and ventral epidermis. In the head the boundary is displaced ventrally (the distal border being covered by dorsal epidermis), whereas in the tail it is displaced dorsally (the distal border being covered by ventral epidermis). After amputation the formation of the wound epithelium differs according to the orientation of the amputation surface. On an anterior transverse surface (which will give rise to a head) it is the dorsal epidermis that stretches to form the wound epithelium, thus reestablishing the topography of a cephalic end. Conversely, on a posterior transverse surface it is the ventral epidermis that covers the wound, thus mimicking the topography of a caudal end. On longitudinal section surfaces both the dorsal and the ventral epidermis stretch to cover the wound and their suture coincides with the dorsal/ventral 'hinge', as in the intact lateral border. Once wound healing is completed intercalary regeneration starts and the missing levels appear in basipetal sequence (fig. 31), each of them being determined before the morphological structure of the preceding level becomes apparent [*Chandebois,* 1973, 1976 a, c].

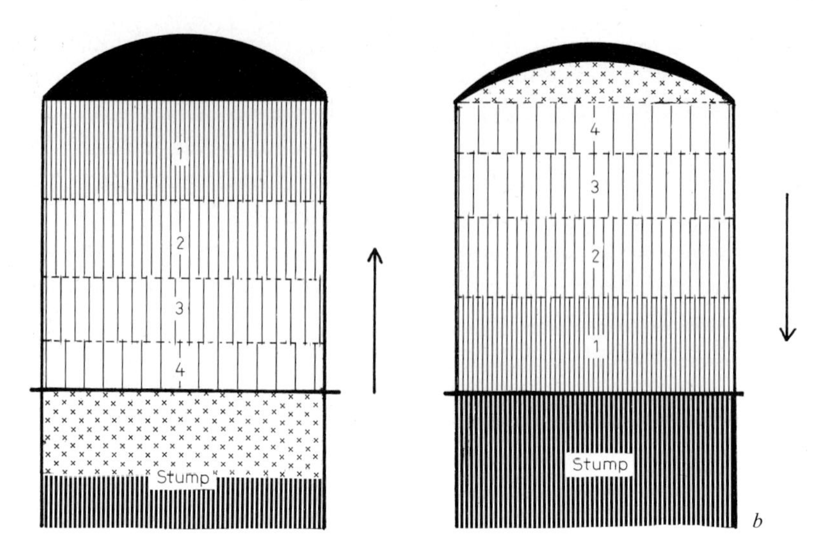

Fig. 31. Comparison between basipetal *(a)* and basifugal *(b)* regeneration. Crosses = Zone where the cells of the regenerate are produced; black = distalmost level, which is re-established during wound closure; arrows = direction of growth of blastema; 1–4 = sequence of determination and differentiation of the various levels (opposite to the direction of growth of blastema).

In certain forms the determination of the blastema occurs not in basipetal but in basifugal sequence (fig. 31), but this does not mean that the principles are fundamentally different [*Chandebois,* 1973]. The only difference is that in this case the dedifferentiated cells multiply not in the stump but in the most distal part of the blastema, once wound healing is completed. The first cells to be produced remain in contact with the stump and are determined by it to form the most proximal level of the regenerate. A case in point is the regenerating urodele limb. Here the wound epidermis forms a so-called 'apical cap', which is comparable to the apical ectodermal ridge of the embryonic limb bud (p. 79) in that it maintains beneath it an apical proliferation zone in the mesenchyme [*Faber,* 1971].

Contrary to intercalary regeneration in the strict sense, terminal regeneration is of an *aleatory* nature because the wound healing procedure is subject to chance errors. In certain cases regeneration is impossible, for instance when for some reason the dorsal and ventral epidermis fail to join. In other cases the regenerate does not reproduce the end that was removed. This phenomenon is called 'heteromorphosis'. Thus, in planarians one frequently obtains a head instead of a tail and conversely.

Although we now begin to see how a repair programme is established in the adult organism, we still know next to nothing of the epigenetic phenomena involved in the structuring of the regenerate. In the case of urodele limb regeneration the blastema shows many properties which parallel those of the embryonic limb bud [*Faber*, 1971]. The mesenchyme produced by the dedifferentiation of bone, cartilage, muscle tissue and connective tissue reproduces the missing levels in proximo-distal sequence under the influence of the apical cap formed by the wound epidermis. If the cap is removed and grafted to the base of the blastema a reduplicated regenerate is obtained, because the original cap reconstitutes itself [*Thornton and Thornton*, 1965]. In no other regenerating system has it been possible to carry the comparison with embryonic development so far, because the experimental data are lacking. We only know that the determination of certain rudiments in the regenerate requires an inductive influence emanating from another rudiment that was determined earlier. For instance, the regenerated brain induces the eyes in planarians [*Lender*, 1952] and the olfactory grooves in nemerteans [*Sandoz*, 1965]. All this shows that *the repair programme does not assign a specific destiny to every cell in the blastema. All it does is to start the cells on a chain of readjustments, ultimately resulting in the definitive structuring of the regenerate.*

A fundamental fact we have demonstrated above (p. 66) is that one and the same cell is involved in several different features of the same structural pattern through various different aspects of its individuality, each aspect being determined by a different factor. The programme of a readjustment therefore comprises several different 'prepatterns'; these are established on the basis of different sources of information which can be dissociated experimentally. The various features of the pattern that are determined by those different prepatterns interfere with each other. It is self-evident that the programming of repair in adult organisms must be based on the same principles, and indeed the conclusions drawn from early and recent experiments on planarians point in this direction. The emergence of the blastema is not brought about simply by the accumulation of dedifferentiated cells. To be incorporated into the blastema the cells must be integrated into a particular level that is determined by the level that was laid down before it (fig. 31). There are no morphological criteria to distinguish the various levels; only experiments can reveal their existence [*Chandebois*, 1957]. It is important to realise, however, that this level-specific determination is not sufficient to make reappear the visible features characterising each level in the normal adult. A regenerate formed on a stump whose tissues present a topography that is different from normal can attain an almost normal length without being able to organise itself completely [*Chandebois,* unpublished results]. Such a regenerate may for instance be identifiable as a head on the basis of its movements and the presence in it of an unpaired intestinal coecum, while remaining without eyes and auricles. It thus seems that distalisation, although it plays a determining role, does not programme the entire regeneration process, as has been thought until now. Apparently all it does is to reconstitute the levels of a covert graded pattern by conferring on the cells certain invisible features of their individualities. It thereby plays an indispensable part in the emergence of the blastema. However, at the time they are integrated into a newly appearing level of the blastema the cells are necessarily located close to the differentiated stump tissues, and these may determine other features of their individualities, particularly their forming part of a specific tissue (fig. 32). Thus, other organisational features appear in the blastema, which interfere with the covert graded pattern and later become progressively more complex through successive readjustments. For instance, dedifferentiated cells would be neuralised in contact with the ends of the nerve cords and would then organise themselves as brain because they are integrated into a particular cephalic level of the covert graded pattern. The brain would then induce the sensory organs at a later stage (fig. 32 B_{1-3}). *This concept explains the determining role of the stump in the organisation of the regenerate, as well as the fact that the stump is unable to correct for anomalies caused by errors in the distalisation procedure.*

In the case of basipetal regeneration, if the amount of cells is insufficient to allow all missing levels to be reformed by epimorphosis, intercalary regeneration proceeds by a shift in the cellular equilibrium in the stump, i.e. by morphallaxis, without correcting for any anomalies that may have arisen during epimorphosis. For instance, in marine planarians it sometimes happens that as a result of partial failure of wound closure a half head forms on the left or right half of the section surface [*Chandebois,* 1957, 1976c]. In this case

---

Fig. 32. Diagrammatic representation of interference phenomena during the differentiation of a planarian regenerate. A_{1-3}, reconstitution of the various levels of the covert graded pattern in basipetal sequence (cf. fig. 31a). B_{1-3}, determination of the features of the overt pattern under the influence of the stump (large arrows) as it interferes with the reconstitution of the graded pattern. Readjustments (e.g. inductive events, small arrows) complete the final pattern. (For simplicity the various levels of the overt pattern are indicated by their fully differentiated structures, although in reality at the time the levels are specified they are only determined, not yet visible).

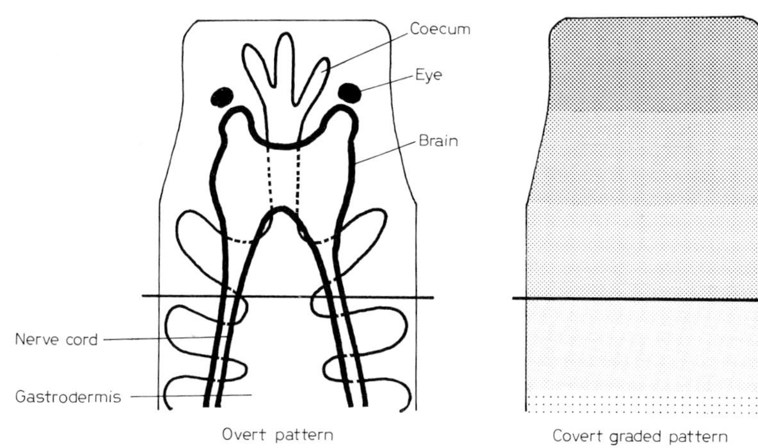

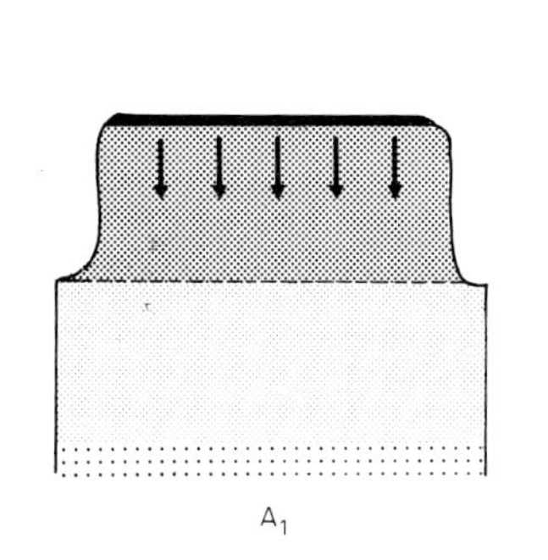

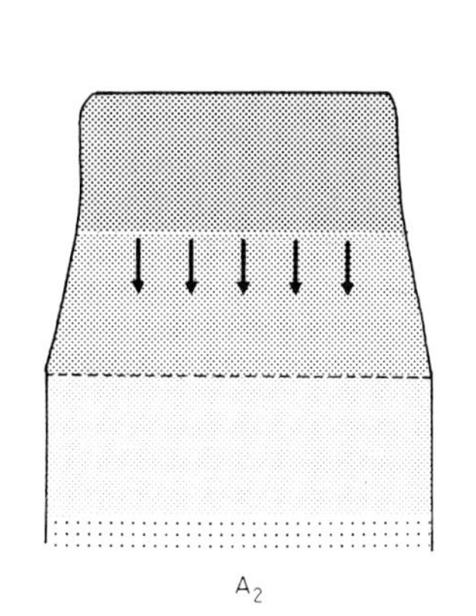

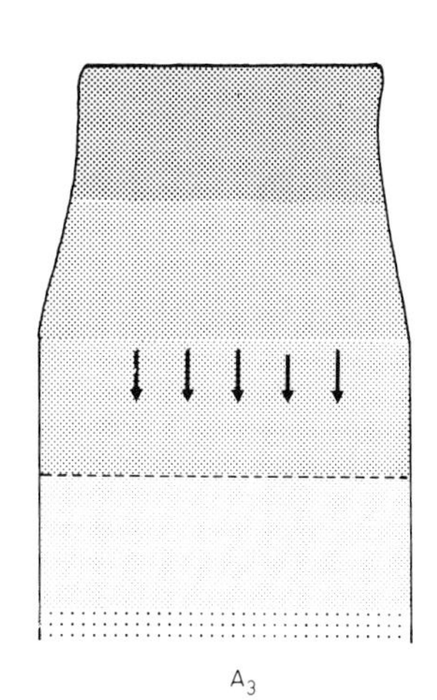

A_1 A_2 A_3

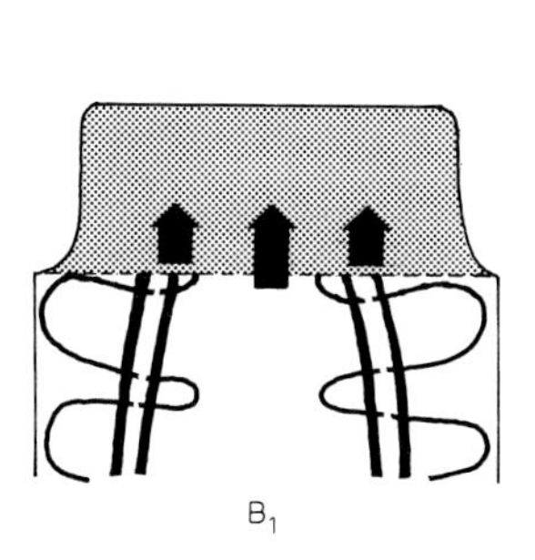

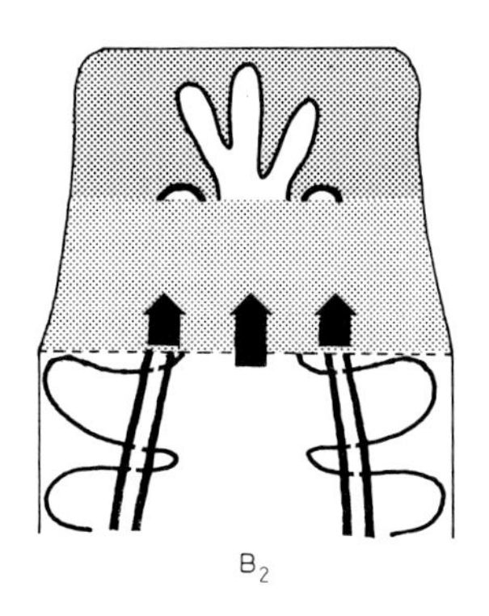

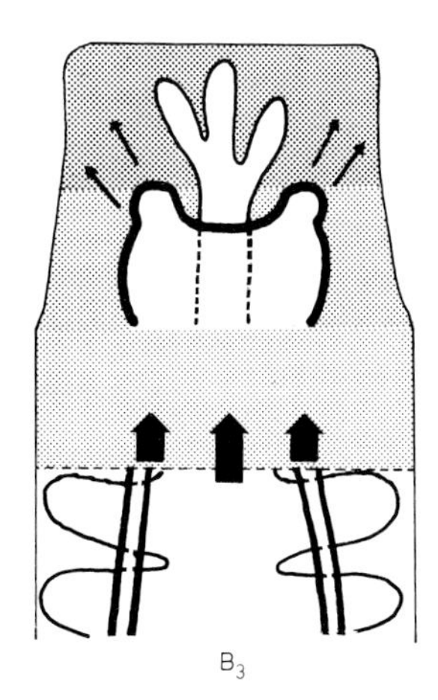

B_1 B_2 B_3

morphallaxis is only observed in that part of the stump that has produced the asymmetrical regenerate. When a heteromorphic tail is formed on a prepharyngeal section surface it leads to the appearance in the stump of a pharyngeal zone of inverted polarity.

Notwithstanding the basic differences on which the distinction between embryonic regulation and regeneration is based, the two phenomena present certain analogies. It is important to stress these at the end of this chapter. We can say that *in both cases the immediate effect of an experimental interference is a change in the elementary social behaviour of the cells in the region affected by the operation.* The cells rearrange themselves, they may dedifferentiate, and they increase their mitotic activity, which enables them to reconstitute (regulation) or to establish (regeneration) the programme for the first readjustment(s) that occur subsequent to the operation. In both cases the features of organisation that were acquired previously also play a part, because they cannot be erased, nor remodelled immediately.

Retrospect and Prospect

We have compared the pluricellular animal to a human society. In such a society each individual acquires knowledge from his contemporaries, which then comes to fruition thanks to the heritage he has received from his ancestors – in this way the individual is integrated into a particular civilisation and participates in its further progress. It must be said that the present-day concepts of morphogenesis conjure up an entirely different sort of society, which one could meaningfully compare to a society ruled in absolute, totalitarian fashion. All individuals are alienated from themselves and placed in a system set up nobody knows precisely how or by whom, a system that at all times forces an individuality upon them and thereby fixes in minute detail the norms of the society and the course of its further progress. In fact it is generally thought that the origin and maintenance of animal organisation require the presence of a pre-existing, invisible organisation which, since *Stern* [1954] coined the term, is usually called a 'prepattern'. In the literature, therefore, this term has a meaning that is very different from the one we have given it in this book. On the one hand, it tacitly assumes that there is only a single source for the information that calls into being or modifies the features of a pattern. On the other hand, it leaves the way open for all sorts of speculation as to the nature, the properties and the origin of this invisible organisation. For most authors this 'blueprint' is some sort of gradient, and that is why we will here use the term 'gradient-prepattern' to avoid all possible confusion.

These ideas, which already became very popular after *Child* [1941] published his well-known concept of 'metabolic gradients', saw a revival when *Wolpert* [1969, 1971] reformulated them in his concept of 'positional information', no doubt because in this concept the problems of morphogenesis are reduced to very simple principles offering to mathematical biologists a range of possibilities for theoretical research. A gradient would be set up between two 'reference points' which have different 'boundary values'. In a way which remains to be discovered these reference points would specify the 'positional information', which in its turn would be 'interpreted' in different ways by the cells, depending on the number of cells separating each of them from the reference points. Restitution and regeneration are explained by assuming that the level where the gradient is interrupted by the operation re-establishes the boundary value of that end of the gradient that has been removed. Although hallowed by frequent use, this notion of the 'gradient-prepattern' remains elusive. In particular, it is difficult to see how the extreme diversity of patterns seen in pluricellular organisms could fundamentally originate from the diversity in the slopes of simple gradients. And it is even more difficult to envisage the epigenetic mechanism by which gradients would be transformed and diversified in the course of ontogenesis. (The term 'epigenetic' is used here in the sense of the de novo formation of structure, as opposed to 'preformistic'; the term also has another, more recent meaning when used for processes occurring outside the genome, as opposed to 'genetic'; the latter usage stems from the erroneous notion that the DNA contains a 'blueprint for development'.) If one evaluates the sheer abundance and the significance of the facts that cannot find a place within these concepts, notably embryonic induction [*Wolpert,* 1969], even without taking into account those facts that are in formal contradiction to them [*Chandebois,* 1976b], one is astonished by the fact that the notions of 'gradient', 'prepattern', 'positional value', etc. are still so widely used today.

If we disregard the mistakes made in reformulating the 'classical' theories without calling them into question first, and in attempting to transform them in terms of an excessive reductionism, the theories themselves are not without merit. If, knowing what we know today, we retrace the history of embryology, we find that in retrospect the old embryologists often passed very close by the truth. One also discovers that very curious historical coincidences were responsible for the fact that experimental research carried on with rigour and objectivity during many decades has ultimately nevertheless resulted in a science that is so to speak 'mythical' in character and incapable of accommodating many of the most solid experimental findings.

The idea of explaining the emergence of visible structure on the basis of a pre-existing invisible organisation has its roots in the old theory of preformation, and the history of this theory in its turn is indissolubly bound up with that of the concepts based on the notion of 'gradient-prepattern'. The introduction of the theory of preformation, going back to the 17th century, was not due to observation but to purely philosophical thinking. In 1661 *Harvey* had discovered the mammalian germ in the uterus. Having searched in vain for organisational features in it that might foreshadow the features of the adult, he went back to the old ideas of *Aristotle* saying that in each new ontogenesis everything emerges de novo, and that the various parts of the embryo are interdependent in their development. Today, we can see how true these ideas were. But every thesis calls for an antithesis, and it was not long before *Harvey's* thesis evoked a dramatically opposed, purely theoretical concept, to wit that all organ rudiments are already to be found in the egg and develop independently of one another. To decide between these concepts it was necessary to do experiments on living material and to interfere with the relations between the different parts of the embryo. This was only possible much later, and the first object was the cleaving egg. In 1887 *Chabry* was the first to develop a method to destroy one of the two blastomeres of the 2-cell stage, but chance led him to make an unfortunate choice of material: the ascidians, one of the few groups where regulation is no longer possible after fertilisation. Because a half embryo develops in the same way when isolated as when associated with the other half, the theory of preformation seemed to be verified. When *Driesch* showed in 1891 that a similar operation in sea urchins led to normal development of the isolated half, knowledge was not sufficiently advanced for the realisation that the differences between mosaic and regulative eggs are essentially of a temporal nature. The theory of preformation still appeared to be well-founded, though applicable to certain species only.

Spemann's discovery of the induction of the amphibian eye lens in 1904 and of the role of the organisation centre in the amphibian embryo in 1918 showed unequivocally that the theory of epigenesis was well-founded, and laid the foundation for modern embryology. His observations, which were confirmed and rethought by *Waddington* [1932, 1956], led to the introduction of such still valid notions as histogenetic potential, competence and determination. Unfortunately, the sudden and considerable increase in complexity of organisation succeeding neural induction still eluded explanation except on the basis of a pre-existing invisible structure in the inducer or in the induced ectoderm. The major difficulty, therefore, was to account for embryonic regulation, which at that time was considered as a reorganisation phenomenon entirely comparable to regeneration [*Guyénot,* 1927]. From that time on the ideas developed further in paradoxical fashion. The preformation theory, rather than being demolished by the discovery of induction and regulation, regained importance and formed a sort of hybrid with the theory of epigenesis. The emergence and structuring of the embryonic anlagen was attributed to the existence of a pre-established invisible organisation, the fundamental property of which is that it is auto-regulatory. By way of ratiocinations and fragmentary observations the embryologists finally came to believe that they had actually demonstrated the existence of this invisible organisation, which they called the 'morphogenetic field'.

The behaviour of embryonic anlagen after partial removal or transplantation reveals properties which are curiously reminiscent of those of physical fields [*Weiss,* 1939; *Waddington,* 1956]. A particular part of the anlage seems to act as a 'centre' from which the 'intensity' of the 'field' declines gradually, which is reflected in the presence of 'equipotential zones' in the anlage. These zones can be re-established after a defect operation and can be shifted when the centre is displaced, or reduplicated when a second centre is added. However well-chosen and fruitful the analogy with physical fields may have been, the concept of the 'morphogenetic field' has nevertheless so to speak thrown our ideas off the track. Because the analogy was pushed too far the morphogenetic field was considered as a physical but elusive reality. It is interesting to recall how *Weiss* [1939] 'raised the term to the dignity of an object of research'. He starts with an excellent rendering of its origins: 'The *field concept* is an abstraction trying to give expression to a group of phenomena observed in living systems. Essentially it is but an *abbreviated formulation of what we have observed*' (*Weiss'* italics). But he then accedes to the idea that the field must have real existence: 'Only secondarily... has the attempt been made to inject *physical* sense into *symbolic* term... The fact that practically all developmental phenomena exhibit field-like characters in one or the other respect is, indeed, a strong indication that the field concept is not only a circumlocution, but an expression of *physical reality*' (*Weiss'* italics). At that point only biochemical analysis seemed capable of advancing the solution of the problem. However, already in 1956 *Waddington* evaluated the risks of this undertaking as follows: 'It seems unlikely that we can hope to obtain anything like a satisfactory understanding of development in biochemical terms until we can comprehend the whole working of the cells, as regards maintenance as well as change... it is premature to look to biochemistry to provide the main framework of ideas for embryology.'

The fears expressed by *Waddington* [1956] turned out to have been lucid. Thanks to the technique of vital staining under anaerobic conditions metabolic gradients were found wherever one looked for them. One could not have hoped for more. However, what one took for 'morphogenetic fields' and later for universal 'gradient-prepatterns' are nothing but, so to speak, the 'ghosts' of the covert graded patterns. These patterns, which are universal motifs in organisms, are established in the same way as overt patterns, i.e. as a result of the interplay of cell interactions in the course of readjustments; the only difference is in the cell properties and the sources of the information used. The idea that the gradient would be a pre-existing 'blueprint' underlying any visible structure was the more credible as covert graded patterns do indeed often serve as 'prepatterns' during the programming of readjustments (antecedent or imprinting prepatterns).

By a curious coincidence the renewed interest these classical theories of morphogenesis have known during the last decade, notwithstanding their shortcomings and their lack of plausibility, owes much to the vicissitudes to which molecular biology was subjected during that same period: molecular biology, which was also heir to a past where knowledge had to be supplemented by intuition, and was equally restricted by the limited precision of its techniques and the impossibility of knowing exactly what one is doing.

Among the various contributions of molecular biology to the science of development the one that was to have the strongest impact was the scheme of gene regulation proposed by *Jacob and Monod* [1961]. The application of this scheme, originally worked out for bacteria, to eukaryotic cells put an end to some old dilemmas: the capacity of cells containing the same genome to maintain different biosynthetic processes, and the reversibility of the differentiated state. Unfortunately, however, this scheme, which now seems erroneous in several respects, and which reduces all control of protein synthesis to a switching on or off of transcription, contained no elements that could have corrected the misapprehensions implicit in the concepts into which it was introduced, and therefore only contributed to their continued credibility.

Thus, *Mendel's* way of thinking was perpetuated by this belief that the genetic material exerts instant control over three-dimensional organisation, and the discovery of the quasi-immediate translation of mRNA pointed in the same direction. The embryologists accordingly interpreted the increase in complexity that accompanies development as the visible manifestation of the sequential derepression of the genes. And it was not long before experimental data seemed to make this assumption into a certainty. In reality, however, a certain kind of mRNA (just as any other substance) is not detectable below a certain concentration, which means that in development, and using techniques that are not very sensitive, it is not detected prior to a certain stage. Thus, the corresponding gene seems to resume its function only at that stage, while in reality it already does so – together with all other structural genes – from the start of cleavage.

For the same reasons the emergence of new cell types after an inductive event seemed necessarily to be linked up with a concomitant derepression of certain genes, in other words, with the appearance of the corresponding 'effectors'. Earlier, the notion of 'inductive principle' cautiously used by *Spemann* had already been replaced by the idea that an 'inductive substance' passes from the inducer to the induced tissue. That this substance would be the effector of one or more genes (or of a product or products involved in its synthetic activity) seemed to be beyond doubt. From then on induction became the model of choice for the study of gene derepression. The succeeding accumulation of failures is not surprising. In fact the almost exclusive attention paid to the DNA, to which all regulatory functions in the cell were ascribed, eclipsed the phenomenon of competence, to which the early experimental embryologists had already attributed both the capacity to respond to induction and the specificity of the response, and which in some experimental situations renders the actual presence of the inducer unnecessary.

All research on the biochemical aspects of induction seemed to reinforce the idea that every cell responds individually to the action of the inducer. This has helped to postpone the solution of the old problem of the increasing complexity of structure following induction, a solution that depends entirely on our knowledge of the social behaviour of the induced cells. For this reason *Waddington* [1956] had to introduce a sharp distinction between what he called 'evocation' – by which a new tissue type is called into being by an inducer – and 'individuation' – the emergence of organisation within the rudiment formed by the tissue in question. He in fact attributed different causes to these two phenomena: the transfer of an inductive substance for the former, and the activity of a morphogenetic field for the latter. Probably this view has contributed much to the present-day division between experimental embryology (the prime concern of which is with intercellular relations in the genesis of tissues) and theoretical investigations (which are centred on the study of 'gradient-prepatterns' with the aim of explaining the genesis of patterns).

The crucial problem in this latter area was to know how, by way of sequential gene derepression, the linear sequence of nucleotides is first converted into a complex three-dimensional organisation and then maintains this same organisation. Without the notion of 'cytoplasmic memory' the only solution could still be a permanent control on the supracellular level, which forced upon each individual cell the traits of its individuality. The plausibility of the notion of 'gradient-prepattern' could only be enhanced by such reasoning. When the notion was accepted without suspicion by the theorists, all that remained to be done was to conjecture as to which molecular activities could underlie the formation of gradients. Among the best-known possibilities suggested one may mention free diffusion of a 'morphogen' produced at one end of the gradient and used up or broken down at a different rate at the other end [*Wolpert,* 1971]; a phase difference between two signals emitted by the same cell, which functions as a pacemaker [*Goodwin and Cohen,* 1969]; and the transmission of positional information via cell contacts [*MacMahon,* 1973].

This rethinking of the gradient concept opened up apparently promising and certainly exciting new avenues. In fact it allowed the mind to invent its own universe by blending mathematical and philosophical reasoning with molecular biology, the approach to biological problems that has become the most prestigious one. It could not but bring with it a certain disdain, more or less explicitly stated, for the kind of morphological research that still is in the great tradition of embryology. Thus, most theorists have burdened themselves but little with the attainments of the past, nor have most of them grasped the significance of the experimental results published by those who during the past 20 years or so have patiently been unravelling the complicated network of cell interactions. Nonetheless, it is the latter who have given us an astonishingly precise picture of development, and this should compel the molecular biologists to rectify their views, and indeed already suggests to them new working hypotheses.

Thus, throughout the history of embryology all such neglects and errors have converged to construct and maintain the present-day quasi-static view of pattern formation, centred as it is on the constant relation of cells with gradients and of gradients with genes – even while the data were accumulating that now enable us better to understand the dynamics of morphogenesis. *Holtfreter* [1938], having isolated and cultured various prospective areas of the amphibian blastula, already described the phenomenon that we have called autonomous progression. Since the work of *Spemann* and his collaborators we know that cells must be competent to engender a certain tissue under the influence of an inducer. When the rigorous concatenation of inductions in the embryo had been demonstrated, the summation of extracellular information and the existence of a cytoplasmic memory were already evident. We have known for a long time that the elimination of homotypic cell interactions stops autonomous progression. This fact, mentioned in numerous original articles but very rarely, if ever, in the current theories, shows that the activity of a cell population is more than the sum of the activities of the individual cells. Finally, the fact that disaggregation of a cell population definitively obliterates certain of its organisational features should long ago have suggested that cells possess a collective memory.

We can now understand which was the fundamental principle that eluded the embryologists when they tried to establish the principles of morphogenesis. Whether the determination of the germ layers, organogenesis or regeneration are concerned, *the cytoplasmic activities become diversified because the cells are engaged in transformations that proceed automatically and interfere with some previously acquired organisation. This process is always started by a simple trigger, which sets going molecular machineries that are already there and ready to function.* During organogenesis this dynamic process is the progression of differentiation. It is set going by heterotypic actions whose only role is to intensify a luxury metabolic strategy that had already been stimulated previously. Thus, the organisation acquired earlier forces the cells to effect a readjustment, as a result of which the organisational features increase in complexity. During the determination of the future germ layers the dynamic process we spoke of is the GA clock. The trigger is the activation of the egg by the sperm or by any parthenogenetic factor, which destabilises the primers produced in the oocyte. The unequal distribution of cytoplasmic constituents which, save in exceptional cases, do not in themselves play a determining role, calls into being the organisation that is required for this temporal transformation of the egg cytoplasm to result in the diversification of the activities of the blastomeres.

Morgan [1934] had already foreseen the relation of the organisation of the egg with the beginnings of differentiation. His hypothesis, which has been amply verified during the years that followed, is still valid in broad outline: the reason why the cleavage nuclei, when they resume their activity, do not do so all in the same manner, is that they come to lie in different parts of the heterogeneous egg cytoplasm. Later, to satisfy the new requirements set by the introduction of *Jacob and Monod's* [1961] views, one had to assume that the resumption of transcription concerns different genes in different parts of the embryo, and consequently that each prospective area must contain particular 'effectors'. The foundation of this theory appeared unshakeable because there actually exist ooplasms where large granules are concentrated that possess the properties of 'determinants'. Although many authors expressed doubts of various sorts one could still continue speculating about invisible determinants. Thus, *Reverberi* [1972] explains the role of the mitochondria in the determination of mosaic eggs by their association with 'macromolecules capable of activation of single different genes'. Such beliefs have without any doubt distorted the interpretation of so many facts that show very clearly that in early development an 'endodermal programme' is produced first and is then succeeded by an 'ectodermal programme'. In addition, ooplasmic segregation (to which one still attributes the segregation of the effectors) has always seemed to be linked up with a pre-existing three-dimensional organisation. The fact that centrifugation does not prevent normal development suggested that this organisation resides in the cortex, which would possess an invisible texture. The discovery of species where the cortex is indeed structured, and where it contributes to the stabilisation of certain ooplasms, seemed to constitute a strong argument for this.

Now that we have followed the development of these ideas it comes as a surprise to find that the oocyte counts among the most commonplace cells in the organism. The pattern it contains is restricted to two polarity axes. The other features that are sometimes observed in the egg cytoplasm have no role whatever in the programming of development; they are so to speak 'ornaments' resulting from the distribution of substances or organelles. On top of this, the two axes mentioned are usually determined by fortuitous circumstances: if not under the blind influence of the external milieu, then as a consequence of the activities of the mitotic apparatus. In this latter case it is remarkable that we have to do with the most archaic constituent of the 'hardware' of the egg which, without having undergone noticeable changes from its first appearance in the unicellular eukaryotes, directs the maturation divisions, amphimixis and cleavage.

Thanks to a new technique allowing the replacement of the egg nucleus by any other embryonic nucleus, *King and Briggs* [1954] had already demonstrated that the cytoplasm of the egg controls the function of the nucleus; this was later confirmed for the specific functions of transcription and DNA replication [*Gurdon*, 1967]. Strangely enough this spectacular discovery failed to change the views of the theorists. They continued to view the DNA – to which they still generally attributed so to speak transcendental functions in the cell – as especially designed to contain in some hidden form the programme of development, together with the memory of the species. Thus, with the idea that structural patterns are reproduced thanks to a code contained in the genetic material, the old theory of preformation obtruded itself anew, though in a different, more concrete and therefore seemingly more credible guise.

In the joint adventure in which embryologists and molecular biologists are engaged the great loser thus turns out to be the DNA. We now see it reduced to the rank of a 'machine' that cannot operate unless it is manipulated by the cytoplasm – which previously seemed to be completely subjected to its control. Today this manipulation by the cytoplasm presents itself to us as an ingenious strategy. To establish the programme for it (fig. 33, 1) the oocyte lineage registers extracellular information in a certain sequence: first information coming from the follicle cells, to which later is added information furnished by substances carried by the blood, first of all the sexual hormones. However, when this registration of extracellular information starts

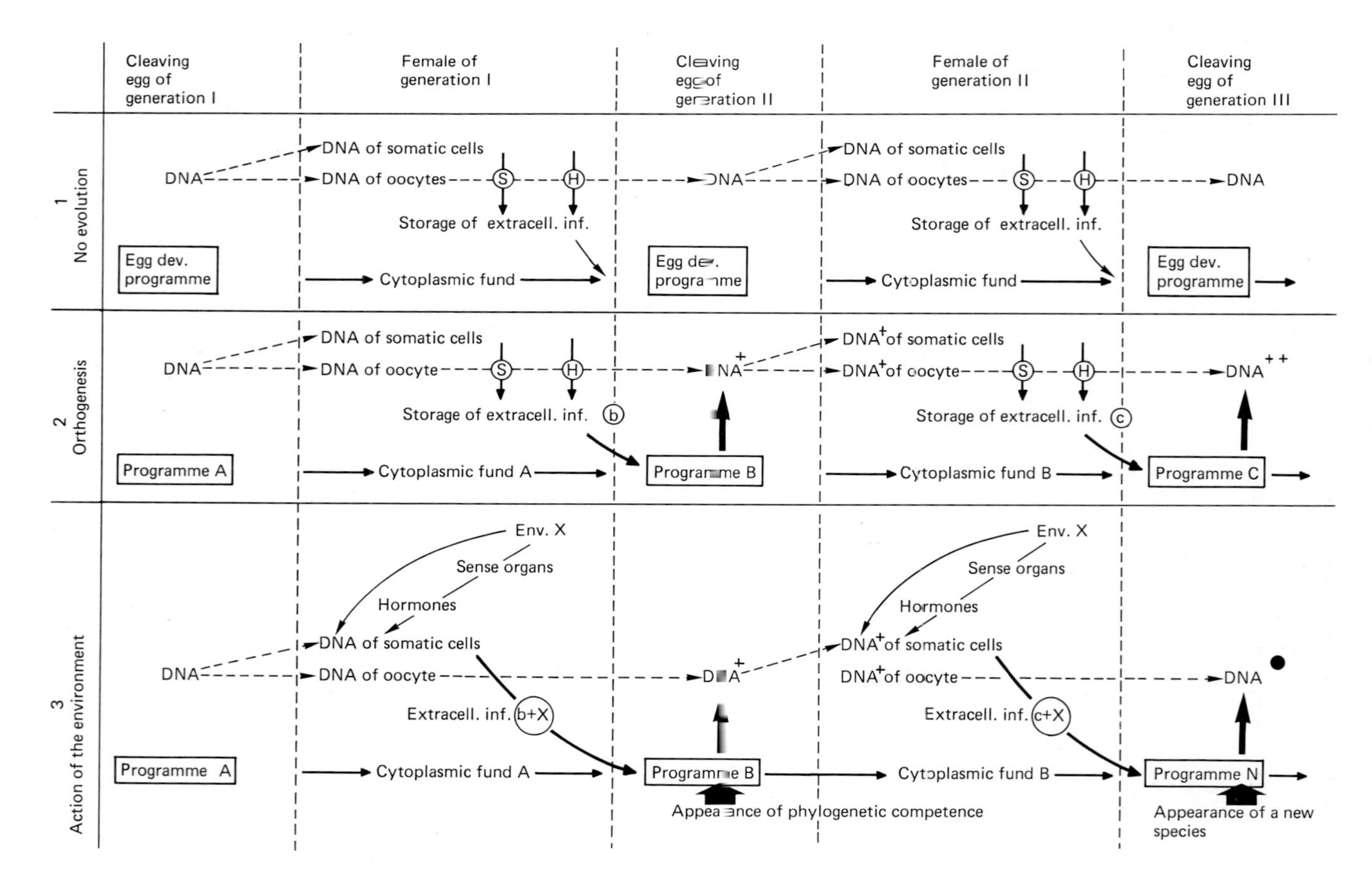

Cleaving egg of generation I
Female of generation I
Cleaving egg of generation II
Female of generation II
Cleaving egg of generation III
1 No evolution
2 Orthogenesis
3 Action of the environment
DNA
DNA of somatic cells
DNA of oocytes
S
H
Storage of extracell. inf.
Cytoplasmic fund
Egg dev. programme
DNA of oocyte
Programme A
Cytoplasmic fund A
Programme B
Cytoplasmic fund B
Programme C
b
c
Env. X
Sense organs
Hormones
Extracell. inf. b+X
Extracell. inf. c+X
Appearance of phylogenetic competence
Programme N
Appearance of a new species

the memory of the oocyte lineage is by no means empty. It 'inherits' cytoplasmic information contained in the egg from which it arises. Very probably certain molecular activities are maintained from one generation to the next: without them the reactivation of the DNA during meiotic prophase would be impossible. These molecular activities represent what we may call the 'cytoplasmic fund' of the species. The old theory of *Weismann*, which has recently been refuted again by *Nieuwkoop and Sutasurya* [1979, 1981] in favour of a purely epigenetic concept of germ cell differentiation, is probably still partially true. If in certain primitive species somatic cells are capable of giving rise to germ cells, this must be because they have retained the cytoplasmic fund of the species, just as somatic cells sometimes retain a latent metabolic strategy that enables them to transdifferentiate.

The proposed existence of a cytoplasmic fund of the species, which links the ascendance to the descendance, makes the individual organism appear as a simple 'cogwheel' among a multitude of other cogwheels making up a more complicated 'machine', the animal world. Just as the individual organism, this machine, which has been running for aeons, is the creator of new forms that become more and more complex and diversified. This raises the question whether the automatism of individual development could not aid in resolving the problem of evolution, just as the automatism of the functioning of the cell holds the key to the understanding of development. Even if we had at our disposal all the space required for a discussion of this vast subject, we would not have the pretension of being capable of it. Here we therefore only want to draw attention to some major points and to propose some ideas that have been suggested to us by the novel conception of development proposed in this book. Otherwise we refer the reader to some books in which the current theories of evolution emerge considerably weakened from the confrontation with their authors' erudition [*Grassé*, 1973; *de Issekutz-Wolsky and Wolsky*, 1976].

If we summarily compare the modalities of development to the workings of a computer, it appears that they are determined at once by the design of

---

Fig. 33. Theoretical scheme to explain organic evolution. *1* Transmission of a cytoplasmic fund from generation to generation. In each oocyte lineage this cytoplasmic fund is used together with stored extracellular information to establish the egg developmental programme. S,H = Information furnished by somatic cells and hormones, respectively. *2* This allows the automatic modification of structural patterns during a phylogenetic lineage ('orthogenesis'). *3* It may also lead to the appearance of periods of phylogenetic competence, during which the environment can alter the programming of development. Further explanation in text.

the 'hardware' (i.e. the genetic code) and by the content of the cytoplasmic memory of the egg (i.e. the cytoplasmic fund of the species plus the information contributed by the maternal somatic cells). Consequently, if the phenotype of an individual differs to some extent from that of its parents the primary cause does not necessarily have to be a genetic mutation (as is still assumed in neo-Darwinism). A cytoplasmic alteration that does not necessarily affect the structure of the DNA can have analogous effects. We already encounter this idea in *Dalcq's* [1957] concept of ontomutations. Nevertheless, it is impossible to assume – as has been done up to now – that any aberration whatever that is due to chance will manifest itself in the emergence of a new form, even if it were the worst monstrosity destined to die early in development. In fact we know that development is impossible when the egg nucleus is replaced by a nucleus of a different species [*Moore,* 1958] or when germ cells differentiate in the gonad of another species [*Reynaud,* 1976]. Even if a sperm is accepted by the egg of another species and successful amphimixis ensues, complete and entirely normal development is impossible. Development is usually blocked at the gastrula stage; at the very best it will lead to a sterile adult. It is clear, therefore, that the developmental programme must 'match the hardware'. When a genetic mutation somehow occurs in a sperm or oocyte lineage, offspring can only be produced if the lineage concerned still belongs to the same species. Therefore, for a new species to appear that is incapable of hybridising with that of its progenitors, two conditions must be fulfilled at once: first, both the maternal and the paternal DNA must undergo the same mutation, and second, the composition of the cytoplasm must be modified in a particular manner, which is difficult to envisage if there is no causal link between the two events. The most probable solution is that the mutation would occur during the first cleavage division under the influence of a cytoplasmic modification that would have occurred during oogenesis. In this way all eggs of one 'batch' would have received the same 'instructions', so that the same mutation would occur in each of them. Because the males and females engendered by this batch would then belong to the same new species they could reproduce among themselves. As a first support for this hypothesis we may adduce the fact that chromosomal aberrations frequently occur during early cleavage upon nuclear transplantation. These are evoked by the cytoplasm when it differs from the nucleus introduced either in developmental age or in taxonomic provenance [*Gallien,* 1974]. The chromosomal changes we envisage might of course be of a much more subtle nature.

One of the major difficulties *Darwin* and the neo-Darwinists were up against was the suddenness with which new major animal groups have emerged during evolution. For each of these events an incalculable number of mutations would have been required, i.e. an extended sequence of slight transformations, for which there is no fossil record in any of these cases. But what we now know about morphogenesis suggests a plausible explanation. If in a cell population at a certain stage of development the elementary social behaviour of the cells is modified, either as a result of a genetic mutation or under the influence of a non-mutagenic treatment, the visible structural effects will be amplified as the readjustments succeed each other. Bearing this in mind, it now seems legitimate to think that a simple minor change in the cytoplasm of the egg, for instance in its ionic composition, could entirely upset the course of development, resulting in a new plan of organisation for the adult. What is essential here is that the egg should be capable of responding to the change in just this manner, a capacity one could call its 'phylogenetic competence'. This would be comparable to 'ontogenetic competence', the only difference being that it would manifest itself in the egg rather than in a more advanced stage of development. We may remind the reader that we have defined 'ontogenetic competence' as the capacity of certain social species markedly to diversify their adult forms under the influence of factors that have no effect whatever on earlier developmental stages, nor in other species (p. 91).

During ontogenesis the emergence of a particular competence proceeds from autonomous progression – that automatic transformation of the activities of a cell population whose initiation, course and termination are programmed at the time of determination. This leads to the idea that evolution could also have its own prime mover: in this case an automatic transformation of the egg developmental programme from one generation to the next, i.e. in a 'phylogenetic lineage'. In effect, under one of its most remarkable aspects, that of orthogenesis, evolution presents itself as an operation that was programmed in the prototype of the phylum. From generation to generation development is carried further: differential growth is extended, accentuating the hypertely of certain parts of the body while at the same time other parts become rudimentary. The idea of the transmission of a 'cytoplasmic fund' and of its utilisation in the establishment of the egg developmental programme in each new generation allows us to advance an hypothesis to account for this automatic progression of evolution (fig. 33, 2). Let us assume that an egg developmental programme A corresponds to the DNA of a given species. In the absence of any change in the extrachromosomal milieu the

DNA will remain unaltered throughout the phylogenetic lineage, provided programme A is reproduced without any change whatever, that is to say if the oocytes continue registering extracellular information of type a. However, even if the DNA remains unaltered the somatic cells could very well communicate to the oocytes of the second generation a slightly altered type of information, which we will call information of type b. This will interact with the cytoplasmic fund A, and as a result a slightly different programme B will be established. This could then bring about a change in the DNA of the egg, as suggested by the nuclear transplantation results discussed above. In the second generation this change in the DNA would lead to a change in the activities of the somatic cells, which as a result would communicate information of type c to the oocytes of the third generation. This will interact with the cytoplasmic fund B to establish programme C, and so on and so forth. This 'disequilibrium' between the nucleus and the cytoplasm could maintain itself in this way until the extinction of the lineage, or it could come to a halt by itself.

As these multiple transformations succeed each other, periods of phylogenetic competence of limited duration could appear, which would be unique in evolutionary history, just as the periods of histogenetic competence are unique in the history of the individual organism. During such a period of phylogenetic competence the interaction between the cytoplasmic fund and the extracellular information registered by the oocyte could result in the establishment of an egg developmental programme that differs profoundly from those of the preceding generations. For that it would be sufficient for the new programme to bring about a simple modification of the elementary social behaviour of the cells, so that one or more readjustments would take a new direction. The repercussions of such a modification for the plan of organisation would be the more fundamental, the earlier the stage that is affected by it. Such a change could just as well occur spontaneously as under the influence of the external milieu (fig. 33, 3). The latter has a regulatory influence on the physiology of the somatic cells, either directly (e.g. temperature or salinity) or by way of the sense organs or the endocrine system: it follows from this that it is necessarily involved in the specification of the extracellular information that is registered by the oocyte. Certain environmental conditions can remain without effect for generations, but if they happen to interact with the cytoplasmic fund of a phylogenetically competent egg, the resulting modification of the egg developmental programme can be such that a new plan of organisation emerges. The scheme suggested here shows that the fact that it is impossible to change species experimentally detracts nothing from the value of the old Lamarckian theory; in fact it reinforces the views of *de Issekutz-Wolsky and Wolsky* [1976], who have reconsidered the theory to do it renewed justice, making use of the attainments of modern biology.

In conclusion, then, we are now led to the view that organic evolution and individual genesis may be based on the same principles of automation, which enable living matter with time to increase the complexity of its organisation and to diversify it. It is true that the exuberance of forms in the living world, their adaptedness to the environment, and the fortuitous nature of mutations lead us a priori to doubt whether this kind of automation can be a reality – for our tendency to equal living things with machines leads us to think that a programme must always dictate a single and inexorable pathway for the process it controls. However, the function of this kind of automation is essentially *to propel molecular machineries through chains of events during which they themselves continuously change.* Only because of that, and in the absence of other than seemingly insignificant triggering factors, any sort of improvisation, one could almost say any kind of fancy becomes possible.

References

Abeloos, M.: Croissance, morphogenèse et évolution. Année biol. *25:* 281–303 (1949).

Abeloos, M.: Régénération postérieure chez *Magalia perarmata.* C. r. hebd. Séanc. Acad. Sci., Paris *230:* 477–478 (1950).

Abrahamsohn, P.A.; Lash, J.W.; Kosher, R.A.; Minor, R.R.: The ubiquitous occurrence of chondroitin sulfates in chick embryos. J. exp. Zool. *194:* 511–518 (1975).

Adelmann, H.B.: Experimental studies on the development of the eye. IV. The effect of the partial and complete excision of the prechordal substrate on the development of the eye of *Amblystoma punctatum.* J. exp. Zool. *75:* 119–227 (1937).

Affara, N.; Daubas, P.: Regulation of a group of abundant mRNA sequences during Friend differentiation. Devl Biol. *12:* 110–125 (1979).

Aimar, C.; Delarue, M.; Vilain, C.: Cytoplasmic regulation of the duration of cleavage in amphibian eggs. J. Embryol. exp. Morph. *64:* 259–274 (1981).

Aizupet (1935): reference unknown, cited in Vorontsova, M.A.; Liosner, L.D.: Asexual propagation and regeneration (Pergamon Press, Oxford 1960).

Ancel, P.; Vintemberger, P.: Recherches sur le déterminisme de la symétrie bilatérale dans l'œuf des amphibiens. Bull. Biol. Fr. Belg. *31:* suppl., pp. 1–182 (1948).

Anderson, D.T.: The development of holometabolous insects; in Counce, Waddington, Developmental systems: insects, vol. 2, pp. 165–242 (Academic Press, London 1973).

Anthony, D.D.; Zeszotek, E.; Goldthwait, D.A.: Initiation by the DNA-dependent RNA polymerase. Proc. natn. Acad. Sci. USA *56:* 1026–1033 (1966).

Apter, M.J.: Cybernetics and development (Pergamon Press, Oxford 1966).

Apter, M.J.; Wolpert, L.: Cybernetics and development. J. theor. Biol. *8:* 244–257 (1965).

Arbib, M.A.: Automata theory in the context of theoretical embryology; in Rosen, Foundations of mathematical biology, vol. 2, pp. 141–215 (Academic Press, New York 1972).

Ashby, W.R.: Requisite variety and its implication for the controls of complex systems. Cybernetica *1:* 83–89 (1958).

Atlan, H.: L'organisation biologique et la théorie de l'information. Actualités scientifiques et industrielles (Hermann, Paris 1972).

Avel, M.: Sur une expérience permettant d'obtenir la régénération de la tête en l'absence certaine de la chaîne nerveuse ventrale ancienne chez les Lombriciens. C. r. hebd. Séanc. Acad. Sci., Paris *194:* 2166–2168 (1932).

Balinsky, B.I.: On the factors determining the size of the lens rudiment in amphibian embryos. J. exp. Zool. *135:* 255–300 (1957).

Ballantine, J.E.M.; Woodland, H.R.; Sturgess, E.A.: Changes in protein synthesis during the development of *Xenopus laevis.* J. Embryol. exp. Morph. *51:* 137–153 (1979).

Bardeen, C.R.: Embryonic and regenerative development in planarians. Biol. Bull. *3:* 262–288 (1902).

Bart, A.: Conditions locales du déclenchement et du développement de la régénération d'une patte chez l'insecte *Carausius morosus* Br. C. r. hebd. Séanc. Acad. Sci., Paris *269:* 473–476 (1969).

Barth, L.G.: Neural differentiation without organizer. J. exp. Zool. *87:* 371–383 (1941).

Barth, L.G.; Barth, L.J.: Ionic regulation of embryonic induction and cell differentiation in *Rana pipiens.* Devl Biol. *39:* 1–22 (1974).

Beermann, W.: Differentiation at the level of the chromosomes. Cell differentiation and morphogenesis. International Lecture Course, Wageningen, pp. 24–54 (North Holland, Amsterdam 19[illegible]).

Beljanski, M.: The regulation of DNA replication and transcription. Expl Biol. Med., vol. 8 (Karger, Basel 1983).

Bell, E.: Information transfer between nucleus and cytoplasm during differentiation. Symp. Soc. exp. Biol. *25:* 127–144 (1971).

Bellairs, R.: The mechanism of somite segmentation of the chick embryo. J. Embryol. exp. Morph. *51:* 227–243 (1979).

Bellairs, R.; Curtis, A.S.G.; Sanders, E.J.: Cell adhesiveness and embryonic differentiation. J. Embryol. exp. Morph. *46:* 207–213 (1978).

Ben-Ze'Ev, A.; Farmer, S.R.; Penman, S.: Protein synthesis requires cell-surface contact while nuclear events respond to cell shape in anchorage-dependent fibroblasts. Cell *21:* 365–372 (1980).

Bertalanffy, L. von: Principles and theory of growth; in Nowinski, Fundamental aspects of normal and malignant growth, pp. 137–259 (Elsevier, Amsterdam 1960).

Bezem, J.J.; Raven, C.P.: Computer simulation of early embryonic development. J. theor. Biol. *54:* 47–61 (1975).

Blackler, A.W.: Contribution to the study of germ cells in the anura. J. Embryol. exp. Morph. *6:* 491–503 (1958).

Boilly, B.; Combaz, A.: Influence de la chaîne nerveuse sur la régénération caudale de *Nereis diversicolor* O.F. Müller (Annélides Polychètes). C. r. hebd. Séanc. Acad. Sci., Paris *271:* 92–95 (1970).

Boilly-Marer, Y.: Néoformation de parapodes surnuméraires par greffe hétérologue de la paroi du corps chez *Nereis pelagica* L. (Annélide Polychète). C. r. hebd. Séanc. Acad. Sci., Paris *272:* 79–82 (1971).

Boterenbrood, E.C.: Organization in aggregates of anterior neural plate cells of *Triturus alpestris.* Proc. Koninkl. Nederl. Akad. Wetenschap. Ser. C. *61:* 470–481 (1958).
Boterenbrood, E.C.: Differentiation in small grafts of the median region of the presumptive prosencephalon. J. Embryol. exp. Morph. *23:* 751–759 (1970).
Boterenbrood, E.C.; Nieuwkoop, P.D.: Self-organization and dependent development of the prosencephalon in amphibians. Symp. Germ Cells Dev., Pallanza 1960, pp. 714–717 (1961).
Boterenbrood, E.C.; Nieuwkoop, P.D.: The formation of the mesoderm in urodelean amphibians. V. Its regional induction by the endoderm. Roux Arch. EntwMech. Org. *173:* 319–332 (1973).
Bounoure, L.: Le sort de la lignée germinale chez la grenouille rousse après l'action des rayons ultraviolets sur le pôle inférieur de l'œuf. C. r. hebd. Séanc. Acad. Sci., Paris *204:* 1837 (1937).
Brachet, J.: Introduction à l'embryologie moléculaire (Masson, Paris 1974).
Brien, P.: Blastogenesis and morphogenesis. Adv. Morphogen. *7:* 151–203 (1968).
Brien, P.: Propos d'un zoologiste: le vivant, épigenèse, évolution épigénétique (Editions Universitaires, Bruxelles 1974).
Britten, R.J.; Davidson, E.H.: Gene regulation for higher cells: a theory. Science *165:* 349–357 (1969).
Brothers, A.J.: Stable nuclear activation dependent on a protein synthesized during oogenesis. Nature, Lond. *260:* 112–115 (1976).
Bull, A.L.: *Bicaudal,* a genetic factor which affects the polarity of the embryo in *Drosophila melanogaster.* J. exp. Zool. *161:* 221–242 (1966).
Bullière, D.: Utilisation de la régénération intercalaire pour l'étude de la détermination cellulaire au cours de la morphogenèse chez *Blabera craniifer* (Insecte Dictyoptère). Devl Biol. *25:* 672–709 (1971).
Burnett, A.L.: Control of polarity and cell differentiation through autoinhibition; a model. Expl Biol. Med., vol. 1, pp. 125–140 (Karger, Basel 1967).
Burns, R.K.: Urogenital system; in Willier, Weiss, Hamburger, Analysis of development, pp. 462–491 (Saunders, London 1955).
Byskov, A.G.: Does the rete ovarii act as a trigger for the onset of meiosis? Nature, Lond. *252:* 396–397 (1974).
Campbell, R.D.: Development of *Hydra* lacking interstitial and nerve cells ('epithelial Hydra'); in Subtelny, Königsberg, Determinants of spatial organization. 37th Symp. of Soc. for Dev. Biol., pp. 267–293 (Academic Press, New York 1979).
Caplan, A.I.; Ordahl, C.P.: Irreversible gene repression model for control of development. Science *201:* 120–130 (1978).
Chandebois, R.: Recherches expérimentales sur la régénération de la Planaire marine *Procerodes lobata.* Bull. Biol. Fr. Belg. *91:* 1–94 (1957).
Chandebois, R.: Cell transformation systems in Planarians; in Kiortsis, Trampusch, Regeneration in animals, pp. 131–142 (North-Holland, Amsterdam 1965).
Chandebois, R.: General mechanisms of regeneration as elucidated by experiments on planarians and amphibians and by a new formulation of the morphogenetic field concept. Acta biotheor. *22:* 2–33 (1973).
Chandebois, R.: Morphogénétique des animaux pluricellulaires (Maloine, Paris 1976a).
Chandebois, R.: Cell sociology: a way of reconsidering the current concepts of morphogenesis. Acta biotheor. *25:* 71–102 (1976b).

Chandebois, R.: Histogenesis and morphogenesis in planarian regeneration. Monogr. devl Biol., vol. 11 (Karger, Basel 1976c).

Chandebois, R.: Cell sociology and the problem of position effect: pattern formation, origin and role of gradients. Acta biotheor. *26:* 203–238 (1977).

Chandebois, R.: The dynamics of wound closure and its role in the programming of planarian regeneration. I. Blastema emergence. Dev. Growth Different. *21:* 195–204 (1979).

Chandebois, R.: Cell sociology and the problem of automation in the development of pluricellular animals. Acta biotheor. *29:* 1–35 (1980a).

Chandebois, R.: The dynamics of wound closure and its role in the programming of planarian regeneration. II. Distalization. Dev. Growth Different. *22:* 693–704 (1980b).

Chandebois, R.: The problem of automation in animal development: confrontation of the concept of cell sociology with biochemical data. Acta biotheor. *30:* 143–169 (1981).

Child, C.M.: Patterns and problems of development (University Press, Chicago 1941).

Chung, H.M.; Malacinski, G.M.: Establishment of the dorsal/ventral polarity of the amphibian embryo: use of ultraviolet irradiation and egg rotation as probes. Devl Biol. *80:* 120–133 (1980).

Church, R.B.; MacCarthy, B.J.: Ribonucleic acid synthesis in regenerating and embryonic liver. II. The synthesis of RNA during embryonic liver development and its relationship to regenerating liver. J. molec. Biol. *23:* 477–486 (1967).

Clark, W.R.; Rutter, W.J.: Levels of regulation during the ontogeny of insulin in the rat embryo. Fed. Proc. *26:* 603 (1967).

Clavert, J.: Symmetrization of the egg of vertebrates. Adv. Morphogen. *2:* 27–60 (1962).

Clayton, R.M.: Regulating factors for lens fibre formation in cell culture. I. Possible requirement for pre-existing levels of crystallin mRNA. Ophthalmic Res. *11:* 324–328 (1979).

Clayton, R.M.: Cellular and molecular aspects of differentiation and transdifferentiation of ocular tissues in vitro; in Yeoman, Truman, Differentiation in vitro, pp. 81–120 (University Press, Cambridge 1982).

Clayton, R.M.; Odeigah, P.G.; Pomerai, D.I. de; Pritchard, D.J.; Thomson, I.; Truman, D.E.S.: Experimental modifications of the quantitative pattern of crystallin synthesis in normal and hyperplastic lens epithelia; in Courtois, Regnault, Biology of the lens epithelial cell in relation to development, ageing and cataract, pp. 123–136 (Inserm, Paris 1976).

Clayton, R.M.; Pomerai, D.I. de; Pritchard, D.J.: Experimental manipulation of alternative pathways of differentiation in cultures of embryonic chick neural retina. Dev. Growth Different. *19:* 319–328 (1977).

Clayton, R.M.; Thomson, I.; Pomerai, D.I. de: Relationship between crystallin mRNA expression in retina cells and their capacity to re-differentiate into lens cells. Nature, Lond. *282:* 628–629 (1979).

Clever, U.: Genaktivitäten in den Riesenchromosomen von *Chironomus tentans* und ihre Beziehungen zur Entwicklung. I. Genaktivierungen durch Ecdyson. Chromosoma *12:* 607–675 (1961).

Collier, J.R.: Morphogenetic significance of biochemical patterns in mosaic embryos; in Weber, The biochemistry of animal development, vol. 1, pp. 203–244 (Acad. Press, New York 1965).

Conklin, E.G.: The development of centrifuged eggs of ascidians. J. exp. Zool. *60:* 2–80 (1931).

Counce, S.J.: The causal analysis of insect embryogenesis; in Counce, Waddington, Developmental systems: insects, vol. 2, pp. 1–156 (Academic Press, London 1973).

Crosby, G.M.: Developmental capabilities of the lateral somatic mesoderm of early chick embryos; PhD thesis, (1967); cited by Stephens et al. (1980).

Czihak, G.: Evidences for inductive properties of the micromere-RNA in sea urchin embryos. Naturwissenschaften *52:* 141–142 (1965).

Dalcq, A.M.: An introduction to general embryology (Oxford University Press, London 1957).

Dalcq, A.M.: Germinal organization and induction phenomena; in Nowinski, Fundamental aspects of normal and malignant growth, pp. 305–494 (Elsevier, Amsterdam 1960).

Dalcq, A.M.: A comparison of the various types of egg organization. Symp. Germ Cells and Development, Pallanza 1960, pp. 704–713 (1961).

Dan, K.: Modified cleavage pattern after suppression of one mitotic division. Expl Cell Res. *72:* 69–73 (1972).

Dan, K.: Unequal division: its cause and significance; in Dirksen, Prescott, Fox, Cell reproduction, pp. 557–561 (Academic Press, New York 1978).

Dan, K.; Nakajima, T.: On the morphology of the mitotic apparatus isolated from echinoderm eggs. Embryologia *3:* 187–200 (1956).

Darnell, J.E.: Implications of RNA-RNA splicing in evolution of eukaryotic cells. Science *202:* 1257–1260 (1978).

Davidson, E.H.: Genomic function in amphibian oogenesis. Atti Simp. Fecondazione, pp. 77–99 (Acad. Naz. Lincei, Roma 1967).

Davidson, E.H.; Hough, B.R.: Synchronous oogenesis in *Engystomops pustulosus,* a neotropic anuran suitable for laboratory studies: localization in the embryo of RNA synthesized at the lampbrush stage. J. exp. Zool. *172:* 25–48 (1969).

Davidson, E.H.; Hough, B.R.: Genetic information in oocyte RNA. J. molec. Biol. *56:* 49–50 (1971).

Davis, L.E.; Burnett, A.L.: A study of growth and cell differentiation in the hepatopancreas of the crayfish. Devl Biol. *10:* 122–153 (1964).

Denis, H.: Activité des gènes au cours du développement embryonnaire (Desoer, Liège 1966).

Dent, J.N.: Survey of amphibian metamorphosis; in Etkin, Gilbert, Metamorphosis, a problem in developmental biology, pp. 271–311 (Appleton Century Crofts, New York 1968).

Deshpande, A.K.; Siddiqui, M.A.Q.: A reexamination of heart muscle differentiation in the postnodal piece of chick blastoderm mediated by exogenous RNA. Devl Biol. *58:* 230–247 (1977).

Dettlaff, T.A.: Cell divisions, duration of interkinetic states and differentiation in early stages of embryonic development. Adv. Morphogen. *3:* 323–362 (1964).

Detwiler, S.R.: On the time of determination of the anteroposterior axis of the forelimb in *Amblystoma.* J. exp. Zool. *64:* 405–414 (1933).

Deuchar, E.M.: Diffusion in embryogenesis. Nature, Lond. *225:* 671 (1970a).

Deuchar, E.M.: Effect of cell number on the type and stability of differentiation in amphibian ectoderm. Expl Cell Res. *59:* 341–343 (1970b).

Deuchar, E.M.: Neural induction and differentiation with minimal numbers of cells. Devl Biol. *22:* 185–199 (1970c).

Devriès, J.: Détermination précoce du développement embryonnaire chez le lombricien *Eisenia foetida.* Bull. Soc. Zool. Fr. *98:* 405–417 (1973a).

Devriès, J.: Aspects du déterminisme embryonnaire au cours des premiers stades de la segmentation chez le lombricien *Eisenia foetida.* Ann. Embryol. Morph. *6:* 95–108 (1973b).

Devriès, J.: Activité génétique au cours de l'embryogenèse de l'oligochète *Eisenia foetida* (autoradiographie à l'uracile-^{3}H et traitement à l'actinomycine D). J. Embryol. exp. Morph. *35:* 403–424 (1976).

Doane, W.W.: Role of hormones in insect development; in Counce, Waddington, Developmental systems: insects, vol. 2, pp. 291–497 (Academic Press, London 1973).
Dohmen, M.R.; Verdonk, N.H.: Cytoplasmic localization in mosaic eggs; in Newth, Balls, Maternal effects in development. 4th Symp. Br. Soc. for Dev. Biol., pp. 127–145 (University Press, Cambridge 1979).
Dollander, A.: Etude des phénomènes de régulation consécutifs à la séparation des deux premiers blastomères de l'œuf de Triton. Archs Biol., Paris *61:* 1–111 (1950).
Dorris, F.: Differentiation of the chick eye in vitro. J. exp. Zool. *78:* 385–415 (1938).
Dym, H.P.; Kennedy, D.S.; Heywood, S.M.: Sub-cellular distribution of the cytoplasmic myosin heavy chain mRNA during myogenesis. Differentiation *12:* 145–155 (1979).
Ebert, J.D.: An analysis of the synthesis and distribution of the contractile protein, myosin, in the development of the heart. Proc. natn. Acad. Sci. USA *39:* 333–344 (1953).
Ebert, J.D.: Annual report of the director of the Department of Embryology, pp. 361–433 (Carnegie Institution, Washington 1959).
Ebert, J.D.; Kaighn, M.E.: Keys to change: factors regulating differentiation; in Locke, Major problems in developmental biology, pp. 20–84 (Academic Press, New York 1966).
Ecker, R.E.; Smith, L.D.: The nature and fate of *Rana pipiens* proteins synthesized during maturation and early cleavage. Devl Biol. *24:* 559–576 (1971).
Ede, D.A.: Control of form and pattern in the vertebrate limb. Control mechanisms of growth and differentiation. Symp. Soc. for exp. Biol., *25:* 235–254 (University Press, Cambridge 1971).
Ede, D.A.: Cell behaviour and embryonic development. Int. J. Neurosci. *3:* 165–174 (1972).
Ede, D.A.; Agerbak, G.S.: Cell adhesion and movement in relation to the developing limb pattern in normal and *talpid*[3] mutant chick embryos. J. Embryol. exp. Morph. *20:* 81–100 (1968).
Ede, D.A.; Flint, O.P.: Patterns of cell division, cell death and chondrogenesis in cultured aggregates of normal and *talpid*[3] mutant chick limb mesenchyme cells. J. Embryol. exp. Morph. *27:* 245–260 (1972).
Ede, D.A.; Flint, O.P.; Wilby, O.K.; Colquhoun, P.: The development of precartilage condensations in limb bud mesenchyme in vivo and in vitro; in Ede, Hinchliffe, Balls, Vertebrate limb and somite morphogenesis. 3rd Symp. Br. Soc. Dev. Biol., pp. 161–180 (University Press, Cambridge 1977).
Ede, D.A.; Kelly, W.A.: Developmental abnormalities in the head region of the *talpid*[3] mutant of the fowl. J. Embryol. exp. Morph. *12:* 161–182 (1964).
Ede, D.A.; Law, J.T.: Computer simulation of vertebrate limb morphogenesis. Nature, Lond. *221:* 224–248 (1969).
Eguchi, G.; Abe, S.I.; Watanabe, K.: Differentiation of lens-like structures from newt iris epithelial cells in vitro. Proc. natn. Acad. Sci. USA *71:* 5052–5056 (1974).
Eguchi, G.; Okada, T.S.: Differentiation of lens tissue from the progeny of chick retinal pigment cells cultured in vitro: a demonstration of a switch of cell types in clonal cell culture. Proc. natn. Acad. Sci. USA *70:* 1495–1499 (1973).
Eguchi, G.; Watanabe, K.: Elicitation of lens formation from the 'ventral iris' epithelium of the newt by a carcinogen, *N*-methyl-*N'*-nitro-*N*-nitrosoguanidine. J. Embryol. exp. Morph. *30:* 63–71 (1973).
Ekblom, P.; Lash, J.W.; Lehtonen, E.; Nordling, S.; Saxén, L.: Inhibition of morphogenetic cell interactions by 6-diazo-5-oxo-norleucine (DON). Expl Cell Res. *121:* 121–126 (1979).
Ellison, M.L.; Ambrose, E.J.; Easty, G.C.: Chondrogenesis in chick embryo somites in vitro. J. Embryol. exp. Morph. *21:* 331–340 (1969).

Ellison, M.L.; Lash, J.W.: Environmental enhancement of in vitro chondrogenesis. Devl Biol. *26:* 486–496 (1971).

Elsdale, T.R.; Jones, K.W.: The independence and interdependence of cells in the amphibian embryo. Cell differentiation. Symp. Soc. exp. Biol., vol. 17, pp. 257–273 (University Press, Cambridge 1963).

Elsdale, T.; Pearson, M.: Somitogenesis in amphibia. II. Origins in early embryogenesis of two factors involved in somite specification. J. Embryol. exp. Morph. *53:* 245–267 (1979).

Etkin, W.: The mechanisms of anuran metamorphosis. I. Thyroxine concentration and the metamorphic pattern. J. exp. Zool. *71:* 317–340 (1935).

Etkin, W.: The acquisition of thyroxine sensitivity by tadpole tissues. Anat. Rec. *108:* 541 (1950).

Eyal-Giladi, H.: Dynamic aspects of neural induction in amphibia (experimentation on *Ambystoma mexicanum* and *Pleurodeles waltlii.* Archs Biol., Paris *65:* 180–259 (1954).

Eyal-Giladi, H.: Differentiation potencies of the young chick blastoderm as revealed by different manipulations. II. Localized damage and hypoblast removal experiments. J. Embryol. exp. Morph. *23:* 739–749 (1970).

Eyal-Giladi, H.; Wolk, M.: The inducing capacities of the primary hypoblast as revealed by transfilter induction studies. Roux Arch. EntwMech. Org. *165:* 226–241 (1970).

Faber, J.: Autonomous morphogenetic activities of the amphibian regeneration blastema; in Kiortsis, Trampusch, Regeneration in animals, pp. 404–419 (North-Holland, Amsterdam 1965).

Faber, J.: Vertebrate limb ontogeny and limb regeneration: morphogenetic parallels. Adv. Morphogen. *9:* 127–147 (1971).

Fallon, J.F.; Saunders, J.W., Jr.: In vitro analysis of the control of cell death in a zone of prospective necrosis from the chick wing bud. Devl Biol. *18:* 553–570 (1968).

Fell, H.B.; Mellanby, E.: Metaplasia produced in cultures of chick ectoderm by high vitamin A. J. Physiol., Lond. *119:* 470–488 (1953).

Flickinger, R.A.: Sequential gene action, protein synthesis and cellular differentiation. Int. Rev. Cytol. *13:* 75–98 (1962).

Flickinger, R.A.: The effect of growth rate on differentiation of chick embryo limb bud mesenchyme in organ culture. Expl Cell Res. *99:* 449–453 (1976a).

Flickinger, R.A.: Effect of rate of replication upon transcription in chick embryo limb bud mesenchyme cells in organ culture. Differentiation *6:* 169–175 (1976b).

Flickinger, R.A.; Stone, G.: Localization of lens antigens in developing frog embryos. Expl Cell Res. *21:* 541–547 (1960).

Franco-Browder, S.J.; Rydt, J. de; Dorfman, A.: The identification of a sulfated mucopolysaccharide in chick embryos, stages 11–23. Proc. natn. Acad. Sci. USA *49:* 643–647 (1963).

Freeman, G.: The role of cleavage in the localization of developmental potential in the ctenophore *Mnemiopsis leidyi.* Devl Biol. *49:* 143–177 (1976).

Freeman, G.: The role of asters in the localization of the factors that specify the apical tuft and the gut of the nemertine *Cerebratulus lacteus.* J. exp. Zool. *206:* 81–108 (1978).

Freeman, G.: The multiple roles which cell division can play in the localization of developmental potential; in Subtelney, Konigsberg, Determinants of spatial organization, pp. 53–76 (Academic Press, New York 1979).

French, V.: Leg regeneration in the cockroach, *Blatella germanica.* II. Regeneration from a noncongruent tibial graft/host junction. J. Embryol. exp. Morph. *35:* 267–301 (1976).

French, V.; Bryant, P.J.; Bryant, S.V.: Pattern regulation in epimorphic fields; cells may make use of a polar coordinate system for assessing their positions in developing organs. Science *193:* 969–981 (1976).

Frenster, J.H.; Herstein, P.R.: RNA in gene expression; in Niu, Segal, The role of RNA in reproduction and development, pp. 330–338 (North-Holland, Amsterdam 1973).

Gallien, L.: Différenciation et organogenèse sexuelle chez les Métazoaires (Masson, Paris 1973).

Gallien, L.: La signification des caryopathies induites dans la transplantation nucléaire chez les Amphibiens. Année biol. *13:* 285–292 (1974).

Gardner, R.L.; Rossant, J.: Determination during embryogenesis; in Elliott, O'Connor, Embryogenesis in mammals, pp. 5–25 (Elsevier, Amsterdam 1976).

Garrod, D.R.; Foreman, D.: Pattern formation in the absence of polarity in *Dictyostelium discoideum.* Nature, Lond. *265:* 144–146 (1977).

Gefter, M.L.: DNA replication. A. Rev. Biochem. *44:* 45–78 (1975).

Gehring, W.J.; Nöthiger, R.: The imaginal discs of *Drosophila;* in Counce, Waddington, Developmental systems: insects, vol. 2, pp. 211–290 (Academic Press, London 1973).

Geilenkirchen, W.L.M.; Timmermans, L.P.M.; Dongen, C.A.M. van; Arnolds, W.J.A.: Symbiosis of bacteria with eggs of *Dentalium* at the vegetal pole. Expl Cell Res. *67:* 477–479 (1971).

Genis-Galvez, J.M.; Castro, J.M.; Battaner, E.: Lens soluble proteins: correlation with the cytological differentiation in the young adult organ of the chick. Nature, Lond. *217:* 652–654 (1968).

Georgiev, G.P.: The structure of transcriptional units in eukaryotic cells; in Moscona, Monroy, Current topics in developmental biology, vol. 7, pp. 1–59 (Academic Press, New York 1972).

Gerhart, J.C., Berking, S.; Cooke, J.; et al.: The cellular basis of morphogenetic changes; in Bonner, Evolution and development, vol. 22, Dahlem Konferenzen, pp. 87–114 (Springer, Berlin 1982).

Gerhart, J.; Ubbels, G.; Black, S.; Hara, K.; Kirschner, M.: A reinvestigation of the role of the grey crescent in axis formation in *Xenopus laevis.* Nature, Lond. *292:* 511–516 (1981).

Gilbert, W.: Starting and stopping sequences for the RNA polymerase; in Losick, Chamberlain, RNA polymerase, pp. 193–206 (Cold Spring Harbor Laboratory, New York 1976).

Goodwin, B.; Cohen, M.H.: A phase-shift model for the spatial and temporal organization of developing systems. J. theor. Biol. *25:* 49–107 (1969).

Graham, C.F.: The design of the mouse blastocyst; in Davies, Balls, Control mechanisms of growth and differentiation. Symp. Soc. exp. Biol., vol. 25, pp. 371–378 (University Press, Cambridge 1971).

Graham, C.F.; Kelly, S.J.: Interactions between embryonic cells during early development in the mouse; in Karkinen-Jääskeläinen, Saxén, Weiss, Cell interactions in differentiation. 6th Sigrid Juselius Fdn Symp., Helsinki, pp. 45–57 (Academic Press, London 1977).

Grassé, P.P.: L'évolution du vivant (Albin Michel, Paris 1973).

Graziosi, G.; Micali, F.: Differential response to ultraviolet irradiation of the polar cytoplasm of *Drosophila* eggs. Roux Arch. EntwMech. Org. *175:* 1–11 (1974).

Grobstein, C.: Epithelial-mesenchymal specificity in the morphogenesis of mouse submandibular rudiments in vitro. J. exp. Zool. *124:* 383–414 (1953).

Grobstein, C.; Holtzer, H.: In vitro studies of cartilage induction in mouse somite mesoderm. J. exp. Zool. *128:* 333–358 (1955).

Grunz, H.: Differentiation of the four animal and the four vegetal blastomeres of the eight-cell-stage of *Triturus alpestris.* Roux Arch. Dev. Biol. *181:* 267–277 (1977).

Grunz, H.; Multier-Lajous, A.M.; Herbst, R.; Arkenberg, B.: The differentiation of isolated amphibian ectoderm with or without treatment with an inducer; a scanning electron microscope study. Roux Arch. Dev. Biol. *178:* 277–284 (1975).

Guerrier, P.: Les caractères de la segmentation et la détermination de la polarité dorsoventrale dans le développement de quelques Spiralia. I. Les formes à premier clivage égal. J. Embryol. exp. Morph. *23:* 611–637 (1970a).

Guerrier, P.: Les caractères de la segmentation et la détermination de la polarité dorsoventrale dans le développement de quelques Spiralia. II. *Sabellaria alveolata* (Annélide, Polychète). J. Embryol. exp. Morph. *23:* 639–665 (1970b).

Guerrier, P.: Les caractères de la segmentation et la détermination de la polarité dorsoventrale dans le développement de quelques Spiralia. III. *Pholas dactylus* et *Spisula subtruncata.* J. Embryol. exp. Morph. *23:* 667–692 (1970c).

Gumpel-Pinot, M.: Ectoderm and mesoderm interactions in the limb bud of the chick embryo studied by transfilter cultures: cartilage differentiation and ultrastructural observations. J. Embryol. exp. Morph. *59:* 157–173 (1980).

Gurdon, J.B.: The transplantation of living cell nuclei. Adv. Morphogen. *4:* 1–43 (1964).

Gurdon, J.B.: Nuclear transplantation and cell differentiation; in De Reuck, Knight, Ciba Fdn Symp. on Cell Differentiation, pp. 65–74 (Churchill, London 1967).

Gurdon, J.B.: Gene expression during cell differentiation; in Heard, Lowenstein, pp. 2–16 (Oxford Biology Readers/University Press, Oxford 1973).

Gustafson, T.: Morphogenetic significance of biochemical patterns in sea urchin embryos; in Weber, The biochemistry of animal development, vol. 1, pp. 139–202 (Academic Press, New York 1965).

Guyénot, E.: Le problème morphogénétique dans la régénération des urodèles: détermination et potentialités des régénérats. Revue suisse Zool. *34:* 127–154 (1927).

Hadorn, E.: Role of genes in developmental processes; in MacElroy, Glass, The chemical basis of development, pp. 779–792 (Hopkins Press, Baltimore 1958).

Hadorn, E.: Konstanz, Wechsel und Typus der Determination und Differenzierung in Zellen aus männlichen Genitalanlagen von *Drosophila melanogaster* nach Dauer-kultur in vivo. Devl Biol. *13:* 424–509 (1966).

Hamburger, V.: Morphogenetic and axial self-differentiation of transplanted limb primordia of 2-day chick embryos. J. exp. Zool. *77:* 379–399 (1938).

Hara, K.: Regional neural differentiation induced by prechordal and presumptive chordal mesoderm in the chick embryo; thesis, Utrecht (1961); cited by Nieuwkoop (1967).

Hara, K.: The cleavage pattern of the axolotl egg studied by cinematography and cell counting. Roux Arch. Dev. Biol. *181:* 73–87 (1977).

Hara, K.; Tydeman, P.; Kirschner, M.: A cytoplasmic clock with the same period as the division cycle in *Xenopus* eggs. Proc. natn. Acad. Sci. USA *77:* 462–466 (1980).

Harrison, R.G.: On relations of symmetry in transplanted limbs. J. exp. Zool. *32:* 1–136 (1921).

Hay, E.D.: Dedifferentiation and metaplasia in vertebrate and invertebrate regeneration; in Ursprung, Results and problems in cell differentiation, vol. 1, pp. 85–108 (Springer, Berlin 1968).

Helff, O.M.: Studies on amphibian metamorphosis. III. The influence of the annular tympanic cartilage on the formation of the tympanic membrane. Physiol. Zool. *1:* 463–495 (1925).

Herkovits, J.; Ubbels, G.A.: The ultrastructure of the dorsal yolk-free cytoplasm and the immediately surrounding cytoplasm in the symmetrized egg of *Xenopus laevis.* J. Embryol. exp. Morph. *51:* 155–164 (1979).

Hillman, N.; Sherman, M.I.; Graham, C.: The effect of spatial arrangement on cell determination during mouse development. J. Embryol. exp. Morph. *28:* 263–278 (1972).

Hogart, P.J.: Biology of reproduction (Blackie, Glasgow 1978).

Holmes, D.S.; Bonner, J.: Sequence composition and organization of the genome and of the nuclear RNA of higher organisms: an approach to understanding gene regulation; in Niu, Segal, The role of RNA in reproduction and development, pp. 304–323 (North-Holland, Amsterdam 1973).

Holtfreter, J.: Eigenschaften und Verbreitung induzierender Stoffe. Naturwissenschaften *21:* 766–770 (1933).

Holtfreter, J.: Veränderungen der Reaktionsweise im alternden isolierten Gastrulaektoderm. Roux Arch. EntwMech. Org. *138:* 163–169 (1938).

Holtzer, H.: An experimental analysis of the development of the spinal column. II. The dispensability of the notochord. J. exp. Zool. *121:* 573–592 (1952).

Hörstadius, S.: Über die Determination im Verlaufe der Eiachse bei Seeigeln. Pubbl. Staz. Zool. Napoli *14:* 251–429 (1935).

Hörstadius, S.: The mechanics of sea urchin development, studied by operative methods. Biol. Rev. *14:* 132–179 (1939).

Hörstadius, S.: Experimental embryology of echinoderms (Clarendon Press, Oxford 1973).

Hsiao, T.H.; Hsiao, C.: Ecdysteroids in the ovary and the egg of the greater wax moth. J. Insect Physiol. *25:* 45–52 (1979).

Humphries, S.; Windass, J.; Williamson, R.: Mouse globin gene expression in erythroid and non-erythroid tissue. Cell *7:* 267–277 (1976).

Huxley, J.: Problems of relative growth (Methuen, London 1932).

Ikenishi, K.; Nieuwkoop, P.D.: Location and ultrastructure of primordial germ cells (PGCs) in *Ambystoma mexicanum.* Dev. Growth Different. *20:* 1–9 (1978).

Illmensee, K.; Mahowald, A.P.: Transplantation of posterior polar plasm in *Drosophila;* induction of germ cells at the anterior pole of the egg. Proc. natn. Acad. Sci. USA *71:* 1016–1020 (1974).

Infante, A.A.; Heilmann, L.J.L.: Distribution of messenger ribonucleic acid in polysomes and non-polysomal particles of sea urchin embryos: translational control of actin synthesis. Biochemistry, N.Y. *20:* 1–8 (1981).

Issekutz-Wolsky, M. de; Wolsky, A.: The mechanism of evolution: a new look at old ideas. Contr. hum. Dev., vol. 4 (Karger, Basel 1976).

Jacob, F.; Monod, J.: On the regulation of gene activity. Cold Spring Harb. Symp. quant. Biol. *26:* 193–211 (1961).

Jacobson, A.G.: The determination and positioning of the nose, lens and ear. I. Interactions within the ectoderm, and between the ectoderm and underlying tissues. II. The role of the endoderm. III. Effects of reversing the anteroposterior axis of epidermis, neural plate and neural fold. J. exp. Zool. *154:* 273–304 (1963).

Jacobson, A.G.: Inductive processes in embryonic development. Science *152:* 25–34 (1966).

Jacobson, A.G.; Gordon, R.: Changes in the shape of the developing vertebrate nervous system analyzed experimentally, mathematically and by computer simulation. J. exp. Zool. *197:* 191–246 (1976).

Jazdowska-Zagrodzinska, B.: Experimental studies on the role of 'polar granules' in the segregation of the pole cells in *Drosophila melanogaster.* J. Embryol. exp. Morph. *16:* 391–399 (1966).

Johnen, A.G.; Albers, B.: The differentiation-pattern in dependence of the proportion between inducing and induced cells studied on gastrula-ectoderm of *Ambystoma mexicanum;* in Saxén, Nordling, Wartiovaara, Control of differentiation. Medical Biology *56:* 317–320 (1978).

Kalthoff, K.: Position of targets and period of competence for UV-induction of the malformation 'double-abdomen' in the egg of *Smittia* spec. (Diptera, Chironomidae). Roux Arch. EntwMech. Org. *168:* 63–84 (1971).

Kandler-Singer, I.; Kalthoff, K.: RNase sensitivity of an anterior morphogenetic determinant in an insect egg (*Smittia* sp., Chironomidae, Diptera). Proc. natn. Acad. Sci. *73:* 3739–3743 (1976).

Kanehisa, T.; Oki, Y.; Ikuta, K.: Partial specificity of low-molecular weight RNA that stimulates RNA synthesis in various tissues. Archs Biochem. Biophys. *165:* 146–152 (1974).

Karkinen-Jääskeläinen, M.: Permissive and directive interactions in lens induction. J. Embryol. exp. Morph. *44:* 167–179 (1978).

Kern, C.H.: Cytochemical and ultrastructural observations on the polar plasm of eggs produced by the female-sterility mutant *fs(1)N* in *Drosophila melanogaster.* Am. Zool. *16:* 189 (1976).

King, T.J.; Briggs, R.: Transplantation of living nuclei of late gastrulae into enucleated eggs of *Rana pipiens.* J. Embryol. exp. Morph. *2:* 73–80 (1954).

Kirschner, M.; Gerhart, J.C.; Hara, K.; Ubbels, G.A.: Initiation of the cell cycle and establishment of bilateral symmetry in *Xenopus* eggs; in Subtelney, Wessels, The cell surface; mediator of developmental processes, pp. 187–215 (Academic Press, New York 1980).

Klein, K.L.; Scott, W.J.; Wilson, J.G.: Aspirin-induced teratogenesis: a unique pattern of cell death and subsequent polydactyly in the rat. J. exp. Zool. *216:* 107–112 (1981).

Kochav, S.; Eyal-Giladi, H.: Bilateral symmetry in chick embryo, determination by gravity. Science *171:* 1027–1029 (1971).

Kosher, R.A.; Savage, M.P.; Walker, K.H.: A gradation of hyaluronate accumulation along the proximodistal axis of the embryonic chick limb bud. J. Embryol. exp. Morph. *63:* 85–98 (1981).

Kosher, R.A.; Searls, R.L.: Sulfated mucopolysaccharide synthesis during the development of *Rana pipiens.* Devl Biol. *32:* 50–68 (1973).

Koshland, D.E., Jr.; Kirtley, M.E.: Protein structure in relation to cell dynamics and differentiation; in Locke, Major problems in developmental biology. 25th Growth Symp., pp. 217–249 (Academic Press, New York 1967).

Kroeger, H.: The induction of new puffing patterns by transplantation of salivary gland nuclei into egg cytoplasm of *Drosophila.* Chromosoma *11:* 129–145 (1960).

Kurihara, K.; Sasaki, N.: Transmission of homoiogenetic induction in presumptive ectoderm of newt embryo. Dev. Growth Different. *23:* 361–369 (1981).

Lallier, R.: Biochemical aspects of animalization and vegetalization in the sea urchin embryo. Adv. Morphogen. *3:* 147–196 (1964).

Lash, J.W.: Tissue interaction and specific metabolic responses: chondrogenetic induction and differentiation; in Locke, Cytodifferentiation and macromolecular syntheses, pp. 235–260 (Academic Press, New York 1963).

Lash, J.W.: Differential behaviour of anterior and posterior embryonic chick somites in vitro. J. exp. Zool. *165:* 47–55 (1967).

Lash, J.W.: Phenotypic expression and differentiation: in vitro chondrogenesis; in Ursprung, Results and problems in cell differentiation, vol. 1, pp. 17–24 (Springer, Berlin 1968).

Laviolette, P.: Etude cytologique et expérimentale de la régénération germinale après castration chez *Arion rufus* L. (Gastéropode Pulmoné). Ann. Sci. Nat. Zool. *11:* 427–535 (1954).

Lehmann, F.E.: Phases of dependent and autonomous morphogenesis in the so-called mosaic-egg of *Tubifex;* in MacElroy, Glass, A symposium on the chemical basis of development, pp. 73–84 (Johns Hopkins, Baltimore 1958).

Lender, T.: Le rôle inducteur du cerveau dans la régénération des yeux d'une planaire d'eau douce. Bull. Biol. Fr. Belg. *56:* 140–215 (1952).

Leussink, J.A.: The spatial distribution of inductive capacities in the neural plate and archenteron roof of urodeles. Neth. J. Zool. *20:* 1–79 (1970).

Levak-Svajger, B.; Svajger, A.: Differentiation in homografts of isolated germ layers of the rat embryo. Archs Sci. Biol. *22:* 25–32 (1970).

Lheureux, E.: Importance des associations de tissus du membre dans le développement des membres surnuméraires induits par déviation du nerf chez le Triton *Pleurodeles waltlii* Michah. J. Embryol. exp. Morph. *38:* 151–173 (1977).

Lindahl, P.E.: Zur experimentellen Analyse der Determination der Dorsoventralachse beim Seeigelkeim. I. Versuche mit gestreckten Eiern. Roux Arch. EntwMech. Org. *127:* 300–322 (1932).

Lopashov, G.V.: Levels in stabilization of cell differentiation and its experimental transformation. Differentiation *9:* 131–137 (1977).

Lopashov, G.V.; Khoperskaya, O.A.: Nature of cell interaction during differentiation of amphibian eye rudiments. Dokl. Akad. Nauk. SSSR *175:* 962–965 (1967).

Lopashov, G.V.; Stroeva, O.G.: Morphogenesis of the vertebrate eye. Adv. Morphogen. *1:* 331–337 (1961).

Lorch, I.J.; Danielli, J.F.; Hörstadius, S.: The effect of enucleation on the development of sea urchin eggs. I. Enucleation of one cell at the 2,4 or 8 cell stage. II. Enucleation of animal or vegetal halves. Expl Cell Res. *4:* 253–274 (1953).

Lumsden, A.G.S.: Pattern formation in the molar dentition of the mouse. J. Biol. buccale *7:* 77–103 (1979).

Lundquist, A.; Emanuelson, H.: Polar granules and pole cells in the embryo of *Calliphora erythrocephala:* ultrastructure and ^{3}H-leucine labelling. J. Embryol. exp. Morph. *57:* 79–93 (1980).

MacAvoy, J.W.: β- and γ-Crystallin synthesis in rat lens epithelium explanted with neural retina. Differentiation *17:* 85–91 (1980).

MacCabe, J.A.; Calandra, A.J.; Parker, B.W.: In vitro analysis of the distribution and nature of a morphogenetic factor in the developing chick wing; in Ede, Hinchliffe, Balls, Vertebrate limb and somite morphogenesis. 3rd Symp. Br. Soc. Dev. Biol., pp. 25–39 (University Press, Cambridge 1977).

MacCabe, J.A.; Saunders, J.W.; Pickett, M.: The control of the anteroposterior and dorsoventral axes in embryonic chick limbs constructed of dissociated and reaggregated limb-bud mesoderm. Devl Biol. *31:* 323–335 (1973).

MacCarthy, M.: Information. A Scientific American book, pp. 1–16 (Freeman, San Francisco 1966).

MacDonald, T.F.; Sachs, H.G.; Orr, C.W.; Ebert, J.D.: External potassium and baby Hamster kidney cells: intracellular ions, ATP, growth, DNA synthesis and membrane potential. Devl Biol. *28:* 290–303 (1972).

MacLaren, A.: Embryogenesis; in Coutinho, Fuchs, Physiology and genetics of reproduction, part B, pp. 297–316 (Plenum Publishing, New York 1974).

MacLoughlin, C.B.: Mesenchymal influences on epithelial differentiation. Symp. Soc. exp. Biol. *17:* 341–357 (1963).

MacMahon, D.: A cell-contact model for cellular position determination in development. Proc. natn. Acad. Sci. USA *70:* 2396–2400 (1973).
Mahowald, A.P.; Allis, C.D.; Karrer, K.M.; Underwood, E.M.; Waring, G.L.: Germ plasm and pole cells of *Drosophila;* in Subtelny, Konigsberg, Determinants of spatial organisation. 37th Symp. Soc. Dev. Biol., pp. 127–146 (Academic Press, New York 1979a).
Mahowald, A.P.; Caulton, J.H.; Gehring, W.J.: Ultrastructural studies of oocytes and embryos derived from female flies carrying the *grandchildless* mutation in *Drosophila subobscura.* Devl Biol. *69:* 118–132 (1979b).
Mahowald, A.P.; Hennen, S.: Ultrastructure of the 'germ plasm' in eggs and embryos of *Rana pipiens.* Devl Biol. *24:* 37–53 (1971).
Mahowald, A.P.; Illmensee, K.; Turner, F.: Interspecific transplantation of polar plasm between *Drosophila* embryos. J. Cell Biol. *70:* 358–373 (1976).
Mangold, O.: Experimente zur Analyse der Determination und Induktion der Medullarplatte. Roux Arch. EntwMech. Org. *117:* 586–696 (1929).
Mangold, O.: Über die Induktionsfähigkeit der verschiedenen Bezirke der Neurula von Urodelen. Naturwissenschaften *21:* 761–766 (1933).
Mangold, O.; Seidel, F.: Homoplastische und heteroplastische Verschmelzung ganzer Tritonkeime. Roux Arch. EntwMech. Org. *111:* 593–666 (1927).
Markert, C.L.; Møller, F.: Multiple forms of enzymes: tissue, ontogenetic and species-specific patterns. Proc. natn. Acad. Sci. USA *45:* 753–763 (1959).
Marzullo, G.; Lash, J.W.: Separation of glycosaminoglycans on thin layers of silica gel. Analyt. Biochem. *18:* 575–578 (1967).
Mauger, A.: Rôle du mésoderme somitique dans le développement du plumage dorsal chez l'embryon de poulet. II. Régionalisation du mésoderme plumigène. J. Embryol. exp. Morph. *28:* 343–366 (1972).
Mayfield, J.E.; Bonner, J.: Tissue differences in rat chromosomal RNA. Proc. natn. Acad. Sci. USA *68:* 2652–2655 (1971).
Mayfield, J.E.; Bonner, J.: A partial sequence of nuclear events in regenerating rat liver. Proc. natn. Acad. Sci. USA *69:* 7–10 (1972).
Medoff, J.: Enzymatic events during cartilage differentiation in the chick embryonic limb bud. Devl Biol. *16:* 118–143 (1967).
Medoff, J.; Zwilling, E.: Appearance of myosin in chick limb bud. Devl Biol. *16:* 138–141 (1972).
Monk, M.: A stem-line model for cellular and chromosomal differentiation in early mouse development. Differentiation *19:* 71–76 (1981).
Moore, J.A.: Transplantation of nuclei between *Rana pipiens* and *Rana sylvatica.* Expl Cell Res. *14:* 532–540 (1958).
Morgan, T.H.: Embryology and genetics (Columbia Univ. Press, 1934).
Morris, G.H.: Teratogenesis; in Newth, Balls, Maternal effects in development. 4th Symp. Br. Soc. Dev. Biol., pp. 351–373 (University Press, Cambridge 1979).
Moscona, A.A.: Environmental factors in experimental studies on histogenesis. La culture organotypique. Associations et dissociations d'organes en culture in vitro. Coll. int. CNRS, Paris, 1961, pp. 155–168.
Moscona, A.A.: Cell recognition, histotypic adhesion and enzyme induction in embryonic cells; in Slavkin, Greulich, Extracellular matrix influences on gene expression, pp. 57–67 (Academic Press, New York 1975).
Müller, U.; Hohl, H.R.: Pattern formation in *Dictyostelium discoideum:* temporal and spatial distribution of prespore vacuoles. Differentiation *1:* 267–276 (1973).

Mulnard, J.: Contribution à la connaissance des enzymes dans l'ontogenèse. Les phosphomonoestérases acide et alcaline dans le développement du rat et de la souris. Archs Biol., Paris *66:* 525–685 (1955).

Nardi, J.B.; Kafatos, F.C.: Polarity and gradients in lepidopteran wing epidermis. I. Changes in graft polarity, form, and cell density accompanying transposition and reorientations. J. Embryol. exp. Morph. *3:* 469–487 (1976).

Nieuwkoop, P.D.: Pattern formation in artificially activated ectoderm (*Rana pipiens* and *Ambystoma punctatum).* Devl Biol. *7:* 255–279 (1963).

Nieuwkoop, P.D.: Problems of embryonic induction and pattern formation in amphibians and birds. Expl Biol. Med., vol. 1, pp. 22–36 (Karger, Basel 1967).

Nieuwkoop, P.D.: The formation of the mesoderm in urodelean amphibians. I. Induction by the endoderm. Roux Arch. EntwMech. Org. *162:* 341–373 (1969a).

Nieuwkoop, P.D.: The formation of the mesoderm in urodelean amphibians. II. The origin of the dorsoventral polarity of the mesoderm. Roux Arch. EntwMech. Org. *163:* 298–315 (1969b).

Nieuwkoop, P.D.: Origin and establishment of embryonic polar axes in amphibian development; in Moscona, Monroy, Current topics in developmental biology, vol. 11, pp. 115–132 (Academic Press, New York 1977).

Nieuwkoop, P.D.; et al.: Activation and organization of the central nervous system in amphibians. J. exp. Zool. *120:* 1–18 (1952).

Nieuwkoop, P.D.; Sutasurya, L.A.: Primordial germ cells in the chordates; embryogenesis and phylogenesis (University Press, Cambridge 1979).

Nieuwkoop, P.D.; Sutasurya, L.A.: Primordial germ cells in the invertebrates; from epigenesis to preformation (University Press, Cambridge 1981).

Niu, M.C.; Deshpande, A.K.: The development of tubular heart in RNA-treated post-nodal pieces of chick blastoderm. J. Embryol. exp. Morph. *29:* 485–501 (1973).

Noda, K.; Kanai, C.: An ultrastructural observation on *Pelmatohydra robusta* at sexual and asexual stages with a special reference to 'germinal plasm'. J. Ultrastruct. Res. *61:* 284–294 (1977).

Nüsslein-Volhard, C.: Maternal effect mutations that alter the spatial coordinates of the embryo of *Drosophila melanogaster;* in Subtelny, Konigsberg, Determinants of spatial organization, pp. 185–211 (Academic Press, New York 1979).

Ohtsu, K.; Naito, K.; Wilt, F.H.: Metabolic basis of visual pigment conversion in metamorphosing *Rana catesbeiana.* Devl Biol. *10:* 216–232 (1964).

Okada, T.S.: 'Transdifferentiation' of cells from chick embryonic eye tissues in culture. Dev. Growth Different. *17:* 289–290 (1975).

Okada, T.S.; Yasuda, K.; Araki, M.; Eguchi, G.: Possible demonstration of multipotential nature of embryonic neural retina by clonal cell culture. Devl Biol. *68:* 600–617 (1979).

Okada, Y.K.; Sugino, H.: Transplantation experiments in *Planaria gonocephala* Dugès. Jap. J. Zool. *3:* 373–439 (1937).

Okada, Y.K.; Takaya, H.: Experimental investigation of regional differences in the inductive capacity of the organizer. Proc. imp. Acad. Tokyo *18:* 505–513 (1942).

Orkin, R.W.; Pollard, T.D.; Hay, E.: SDS gel analysis of muscle proteins in embryonic cells. Devl Biol. *35:* 388–394 (1973).

Ortolani, G.: Cleavage and development of egg fragments in ascidians. Acta embryol. Morph. exp. *1:* 247–272 (1958).

Ortolani, G.; Marino, L.: Sviluppo di quartetti animali isolati dell'uovo delle ascidie dopo trat-

tamento con tripsine ed RNA. Acta Embryol. exp. (gruppo embriologico italiana, 19e reunione), pp. 235–236 (1973).
Parisi, E.; Filosa, S.; Petrocellis, B. de: The pattern of cell division in the early development of the sea urchin *Paracentrotus lividus.* Devl Biol. *65:* 38–49 (1978).
Pasteels, J.: Recherches sur la morphogenèse et le déterminisme des segmentations inégales chez les Spiralia. Archs Anat. microsc. Morph. exp. *30:* 161–197 (1934).
Pasteels, J.: Recherches sur les facteurs initiaux de la morphogenèse chez les Amphibiens anoures. I. Résultats de l'expérience de Schultz et leur interprétation. Archs Biol., Liège *49:* 629–667 (1938).
Pasteels, J.: The morphogenetic role of the cortex of the amphibian egg. Adv. Morphogen. *3:* 363–388 (1964).
Pearson, M.; Elsdale, T.: Somitogenesis in amphibian embryos. I. Experimental evidence for an interaction between two temporal factors in the specification of somite pattern. J. Embryol. exp. Morph. *51:* 27–50 (1979).
Peltrera, A.: Le capacita regolative dell' uovo di *Aplysia limacina* L. studiate con la centrifugazione e con le reazioni vitali. Pubbl. Staz. zool. Napoli *18:* 20–49 (1940).
Perlman, S.M.; Ford, P.J.; Rosbach, M.M.: Presence of tadpole and adult globin RNA sequences in oocytes of *Xenopus laevis.* Proc. natn. Acad. Sci. USA *74:* 3835–3839 (1977).
Perlmann, P.; Vincentiis, M. de: Lens antigen in the microsomal fraction of early chick embryos. Expl Cell Res. *23:* 612–615 (1961).
Petrocellis, B. de; Filosa-Parisi, S.; Monroy, A.: Cell interactions and DNA replication in the sea urchin embryo; in Lash, Burger, Cell and tissue interactions, pp. 269–283 (1977).
Polezhaev, L.V.: Loss and restoration of regenerative capacity in tissues and organs of animals (University Press, Harvard 1972).
Pollard, T.D.; Weihing, T.T.: Actin and myosin and cell movement. CRC crit. Rev. Biochem. *2:* 1–65 (1974).
Pritchard, D.J.; Clayton, R.M.; Pomerai, D.I. de: 'Transdifferentiation' of chicken neural retina into lens and pigment epithelium in culture: controlling influences. J. Embryol. exp. Morph. *48:* 1–21 (1978).
Ranzi, S.: On RNA action in differentiation: induction and differentiation of somites in chick embryo; in MacKinnel, Di Berardino, Blumenfeld, Bergad, Differentiation and neoplasia. Results and problems in cell differentiation, vol. 11, pp. 191–195 (Springer, Berlin 1980).
Rasiló, M.L.; Leikola, A.: Neural induction by previously induced epiblast in avian embryo in vitro. Differentiation *5:* 1–7 (1976).
Raven, C.P.: Oogenesis: the storage of developmental information. Int. Ser. Monogr. on Pure and Applied Biology, Zoology Division, vol. 10 (Pergamon Press, London 1961).
Raven, C.P.: Mechanisms of determination in the development of gastropods. Adv. Morphogen. *3:* 1–32 (1964).
Raven, C.P.; Kloos, J.: Induction by medial and lateral pieces of the archenteron roof, with special reference to the determination of the neural crest. Acta neerl. Morph. norm. path. *5:* 348–362 (1945).
Reinbold, R.: Regulation de l'œil et régénération du cristallin chez l'embryon de poulet opéré en culture in vitro. Archs Anat. microsc. Morph. exp. *47:* 341–358 (1958).
Reverberi, G.: The embryology of ascidians. Adv. Morphogen. *1:* 55–101 (1961).
Reverberi, G.: The fine structure of the ovaric egg of *Dentalium.* Acta Embryol. exp. *1972:* 135–166.

Reverberi, G.; La Spina, R.: Normal larvae obtained from dark fragments of centrifuged *Ciona* eggs. Experientia *15:* 112 (1959).

Reverberi, G.; Minganti, A.: Le potenze dei quartetti animale e vegetative isolati di *Ascidiella aspersa.* Pubbl. Staz. Zool. Napoli *20:* 135–151 (1946a).

Reverberi, G.; Minganti, A.: Fenomeni di evocazione nelle sviluppo dell'uovo di ascidie. Resultati dell' indagine sperimentale sull'uovo di *Ascidiella aspersa* et di *Ascidia malaca* allo stadio di otto blastomeri. Pubbl. Staz. Zool. Napoli *20:* 191–252 (1946b).

Reynaud, G.: Capacités reproductrices et descendance de poulets ayant subi un transfert de cellules germinales primordiales durant la vie embryonnaire. Roux Arch. Dev. Biol. *179:* 85–110 (1976).

Rosen, R.: Some relational cell models: the metabolism repair systems; in Rosen, Foundations of mathematical biology, vol. 2, pp. 217–253 (Academic Press, New York 1972).

Rossant, J.: Investigation of the determinative state of the mouse inner cell mass. I. The fate of isolated inner cell masses transferred to the oviduct. J. Embryol. exp. Morph. *33:* 991–1001 (1975).

Runnström, J.: Plasmabau und Determination bei dem Ei von *Paracentrotus lividus* LK. Roux Arch. EntwMech. Org. *113:* 556–581 (1928).

Rustad, R.C.: Dissociation of the mitotic time-schedule from the micromere 'clock' with X-rays. Acta Embryol. Morph. exp. *3:* 155–158 (1960).

Sala, M.: Distribution of activating and transforming influences in the archenteron roof during the induction of the nervous system in amphibians. I. Distribution in cranio-caudal direction. Proc. Koninkl. Nederl. Akad. Wetenschappen, Ser. C *58:* 635–647 (1955).

Sandoz, H.: Sur la régénération antérieure chez le Némertien *Tetrastemma vittatum.* C. r. hebd. Séanc. Acad. Sci., Paris *260:* 4091–4092 (1965).

Sasaki, N.; Kawakami, I.; Mifune, S.; Tamanoi, I.: The lens induction by a crude protein extracted from rabbit bone-marrow. Mem. Fac. Sci. Kyushu Univ., Ser. E *2:* 159–162 (1957).

Saunders, J.W., Jr.: The proximo-distal sequence of origin of the parts of the chick wing and the role of the ectoderm. J. exp. Zool. *108:* 363–403 (1948).

Saunders, J.W., Jr.: The experimental analysis of chick limb bud development; in Ede, Hinchliffe, Balls, Vertebrate limb and somite morphogenesis. 3rd Symp. Br. Soc. Dev. Biol., pp. 1–24 (University Press, Cambridge 1977).

Saunders, J.W., Jr.; Fallon, J.F.: Cell death in morphogenesis; in Locke, Major problems in developmental biology. 25th Growth Symp., pp. 289–314 (Academic Press, New York 1967).

Saunders, J.W., Jr.; Gasseling, M.T.: Ectodermal-mesenchymal interactions in the origin of limb symmetry; in Fleischmajer, Billingham, Epithelial-mesenchymal interactions, pp. 78–97 (Williams & Wilkins, Baltimore 1968).

Saunders, J.W., Jr.; Gasseling, J.M.; Gfeller, M.D.: Interactions of ectoderm and mesoderm in the origin of axial relationships in the wing of the fowl. J. exp. Zool. *137:* 39–74 (1958).

Sawai, T.: Cycle changes in the cortical layer of non-nucleated fragments of the newt's egg. J. Embryol. exp. Morph. *51:* 183–193 (1979).

Saxén, L.; Koskimies, O.; Lahtii, A.; Miettinen, H.; Rapola, J.; Wartiovaara, J.: Differentiation of kidney mesenchyme in an experimental model system. Adv. Morphogen. *7:* 251–293 (1968).

Saxén, L.; Lehtonen, E.; Karkinen-Jääskeläinen, M.; Nordling, S.; Wartiovaara, J.: Are morphogenetic interactions mediated by transmissible signal substances or through cell contacts? Nature, Lond. *259:* 662–663 (1976).

Saxén, L.; Saksela, E.: Transmission and spread of embryonic induction. II. Exclusion of an assimilatory transmission mechanism in kidney tubule induction. Expl Cell Res. *66:* 369–377 (1971).

Scherrer, K.: Cascade regulation – a model of integrative control of gene expression in eukaryotic cells and organisms; in Kolodny, Eukaryotic gene regulation, pp. 57–129 (CRC Press, 1980).

Schilt, J.: Induction expérimentale d'excroissances par des greffes hétéropolaires chez la Planaire *Dugesia lugubris* O. Schmidt. Ann. Embryol. Morph. *3:* 93–106 (1970).

Searls, R.L.: Isolation of mucopolysaccharide from the precartilaginous embryonic chick limb bud. Proc. Soc. exp. Biol. Med. *118:* 1172–1176 (1965).

Seidel, F.: Untersuchungen über das Bildungsprinzip der Keimanlage im Ei der Libelle *Platycnemis pennipes.* Roux Arch. EntwMech. Org. *119:* 322–440 (1929).

Seilern-Aspang, F.; Kratochwil, K.: Relation between regeneration and tumor growth; in Kiortsis, Trampusch, Regeneration in animals, pp. 452–473 (North-Holland, Amsterdam 1965).

Sengel, P.: The organogenesis and arrangement of cutaneous appendages in birds. Adv. Morphogen. *9:* 181–230 (1971).

Sengel, P.; Mauger, A.: Peridermal cell patterning in the feather-forming skin of the chick embryo. Devl Biol. *51:* 166–171 (1976).

Sengel, P.; Novel, G.: Sur les mécanismes de la morphogenèse du patron plumaire dans la ptéryle spinale de l'embryon de poulet. C. r. hebd. Séanc. Acad. Sci., Paris *271:* 2015–2018 (1970).

Sengel, P.; Rusaouën, M.: Aspects histologiques de la différenciation précoce des ébauches plumaires chez le poulet. C. r. hebd. Séanc. Acad. Sci., Paris *226:* 795–797 (1968).

Signoret, J.: Evidence of the first genetic activity required in axolotl development; in MacKinnell, Di Berardino, Blumenfeld, Bergad, Differentiation and neoplasia. Results and problems in cell differentiation, vol. 11, pp. 71–74 (Springer, Berlin 1980).

Sládecèk, F.: Morphological regulation and swimming capacity in urodela after the reversal of the presumptive brain parts of the neural plate. Folia biol. *6:* 403–419 (1960).

Smith, L.D.: The role of a 'germinal plasm' in the formation of primordial germ cells in *Rana pipiens.* Devl Biol. *14:* 330–347 (1966).

Smith, L.D.; Ecker, R.E.; Subtelny, S.: In vivo induction of physiological maturation in *Rana pipiens* oocytes removed from their ovarian follicles. Devl Biol. *17:* 627–643 (1968).

Smith, L.D.; Williams, M.: Germinal plasm and germinal determinants in anuran amphibians; in Newth, Balls, Maternal effects in development. 4th Symp. Br. Soc. Dev. Biol., pp. 167–197 (University Press, Cambridge 1979).

Smith, L.J.: Embryonic axis orientation in the mouse and its correlation with blastocyst relationships to the uterus. I. Relationship between 82 hours and $4^1/_4$ days. J. Embryol. exp. Morph. *55:* 257–277 (1980).

Sommerville, J.: Gene activity in the lampbrush chromosomes of amphibian oocytes; in Paul, Biochemistry of cell differentiation. II. International review of biochemistry, vol. 15, pp. 79–156 (University Park Press, Baltimore 1977).

Sommerville, J.: Transcription during amphibian oogenesis; in Newth, Balls, Maternal effects in development. 4th Symp. Br. Soc. Dev. Biol., pp. 47–63 (University Press, Cambridge, 1979).

Spemann, H.: Experimentelle Erzeugung zweiköpfiger Embryonen. Sitz. Ber. phys.-med. Ges. Würzburg (1900).

Spemann, H.: Über die Determination der ersten Organanlagen des Amphibienembryo. I–IV. Roux Arch. EntwMech. Org. *43:* 448–555 (1918).

Spemann, H.: Über den Anteil von Implantat und Wirtskeim an der Orientierung und Beschaffenheit der induzierten Embryonalanlage. Roux Arch. EntwMech. Org. *123:* 389–517 (1931).

Spiegel, M.: Protein changes in development. Biol. Bull. *118:* 451–462 (1960).

Stanisstreet, M.; Jumah, H.; Kurais, A.R.: Properties of cells from inverted embryos of *Xenopus laevis* investigated by scanning electron microscopy. Roux Arch. Dev. Biol. *189:* 181–186 (1980).

Stephens, T.D.; Vasan, N.S.; Lash, J.W.: Extracellular matrix synthesis in the chick embryo lateral plate prior to and during limb outgrowth. J. Embryol. exp. Morph. *59:* 71–87 (1980).

Stern, C.: Genes and developmental patterns. Caryologia *6:* suppl., pp. 355–369 (1954).

Sturgess, E.A.; Ballantine, J.E.M.; Woodland, H.R.; Mohun, P.R.; Lane, C.D.; Dimitriadis, G.J.: Actin synthesis during the early development of *Xenopus laevis.* J. Embryol. exp. Morph. *58:* 303–320 (1980).

Sutasurya, L.A.; Nieuwkoop, P.D.: The induction of the primordial germ cells in the urodeles. Roux Arch. EntwMech. Org. *175:* 199–220 (1974).

Takaya, H.: Types of neural tissues induced through the presumptive notochord of newt embryo. Differentiation *7:* 187–192 (1977).

Tanaka, Y.: Effects of the surfactants on the cleavage and further development of the sea urchin embryos. II. The inhibition of micromere formation at the fourth cleavage. Dev. Growth Different. *18:* 113–122 (1976).

Tarkowski, A.K.; Wroblewska, J.: Development of blastomeres of mouse eggs isolated at the 4- and 8-cell stage. J. Embryol. exp. Morph. *18:* 155–180 (1967).

Tata, J.R.: Requirement for RNA and protein synthesis for induced regression of the tadpole tail in organ culture. Devl Biol. *13:* 77–94 (1966).

Thomas, T.L.; Posakony, J.W.; Anderson, D.M.; Britten, R.J.; Davidson, E.H.: Molecular structure of maternal RNA. Chromosoma *84:* 319–335 (1981).

Thomson, I.; Pomerai, D.I. de; Jackson, J.F.; Clayton, R.M.: Lens-specific RNA in cultures of embryonic chick neural retina and pigmented epithelium. Expl Cell Res. *122:* 73–81 (1979).

Thornton, C.S.: Amphibian limb regeneration. Adv. Morphogen. *7:* 205–249 (1968).

Thornton, C.S.; Thornton, M.T.: The regeneration of accessory limb parts following epidermal cap transplantation in urodeles. Experientia *21:* 146 (1965).

Toivonen, S.: Mechanism of primary embryonic induction. Expl Biol. Med., vol. 1, pp. 1–7 (Karger, Basel 1967).

Toivonen, S.; Saxén, L.: The simultaneous inducing action of liver and bone-marrow of the guinea-pig in implantation and explantation experiments with embryos of *Triturus.* Expl Cell Res. *3:* suppl., pp. 346–357 (1955).

Toivonen, S.; Saxén, L.: Late tissue interactions in the segregation of the central nervous system. Ann. Med. exp. Fenn. *44:* 128–130 (1966).

Toivonen, S.; Tarin, D.; Saxén, L.: The transmission of morphogenetic signals from amphibian mesoderm to ectoderm in primary induction. Differentiation *5:* 49–55 (1976).

Townes, P.L.; Holtfreter, J.: Directed movements and selective adhesion of embryonic amphibian cells. J. exp. Zool. *128:* 53–120 (1955).

Trampusch, H.A.L.; Harrebomée, A.E.: Dedifferentiation, a prerequisite of regeneration; in Kiortsis, Trampusch, Regeneration in animals, pp. 341–376 (North-Holland, Amsterdam 1965).

Trinkaus, J.P.: Morphogenetic cell movements; in Locke, Major problems in developmental biology. 25th Growth Symp., pp. 125–176 (Academic Press, New York 1967).

Tsai, S.Y.; Tsai, M.J.; Lin, C.; O'Malley, B.W.: Effect of oestrogen on ovalbumin gene expression in differentiated non-target tissues. Biochemistry, N.Y. *18:* 5726–5731 (1979).

Tung, T.C.; Ku, S.H.; Tung, Y.F.: The development of the ascidian egg centrifuged before fertilization. Biol. Bull. *80:* 153–168 (1941).

Ubbels, G.A.; Bezem, J.J.; Raven, C.P.: Analysis of follicle cell patterns in dextral and sinistral *Limnaea peregra.* J. Embryol. exp. Morph. *21:* 445–466 (1969).

Vignau, J.; Louvet, J.P.; Haget, A.: Premières observations sur un remplacement régulier des embryons abortifs par des ébauches embryonnaires nouvelles, dans les œufs de *Carausius morosus.* Procès-Verb. Soc. Sci. Phys. Nat. Bordeaux *1962:* 89–91.

Vogt, W.: Über regeneratives und regulatives Wachstum. Anat. Anz. *71:* suppl. 141 (1931).

Waddington, C.H.: Experiments on the development of chick and duck embryos cultivated in vitro. Phil. Trans. R. Soc. *221:* 179–230 (1932).

Waddington, C.H.: Genes as evocators in development. Growth suppl. 1. Symp. on Dev. and Growth, pp. 37–45 (1939).

Waddington, C.H.: Principles of embryology (Allen & Unwin, London 1956).

Wallace, H.: Vertebrate limb regeneration (Wiley & Sons, Chichester 1981).

Wartiovaara, J.; Nordling, S.; Lehtonen, E.; Saxén, L.: Transfilter induction of kidney tubules: correlation with cytoplasmic penetration into nucleopore filters. J. Embryol. exp. Morph. *31:* 667–682 (1974).

Weber, R.: Zur Aktivierung der Kathepsine im Schwanzgewebe von *Xenopus*larven bei spontaner und in vitro induzierter Rückbildung. Helv. physiol. pharmacol. Acta *21:* 277–291 (1963).

Weber, R.: Biochemistry of amphibian metamorphosis; in Florkin, Stotz, Morphogenesis, differentiation and development. Comprehensive biochemistry, vol. 28, pp. 145–198 (Elsevier, Amsterdam 1967).

Weijer, C.J.; Nieuwkoop, P.D.; Lindenmayer, A.: A diffusion model for mesoderm induction in amphibian embryos. Acta biotheor. *26:* 164–180 (1977).

Weiss, P.: Principles of development (Holt, New York 1939). Facsimile edition (Hafner, New York 1969).

Whittaker, J.R.: A relationship between increased protein synthesis and loss of melanin synthesis in monolayer cultures of chick retinal pigment cells. Am. Zool. *5:* 80 (1965).

Whittaker, J.R.: Cytoplasmic determinants of tissue differentiation in the ascidian egg; in Subtelny, Konigsberg, Determinants of spatial organization, pp. 29–51 (Academic Press, New York 1979).

Wigglesworth, V.B.: Local and general factors in the development of 'pattern' in *Rhodnius prolixus* (Hemiptera). J. exp. Biol. *17:* 180–200 (1940).

Wilde, C.E.: Factors concerning the degree of cellular differentiation in organotypic and disaggregated tissue cultures. La culture organotypique. Coll. int. CNRS, vol. 101, pp. 183–198 (CNRS, Paris 1961a).

Wilde, C.E.: The differentiation of vertebrate pigment cells. Adv. Morphogen. *1:* 267–300 (1961b).

Wilde, C.E.; Crawford, R.B.: Cellular differentiation in the amniota. III. Effects of actinomycin D and cyanide on the morphogenesis of *Fundulus.* Expl Cell Res. *44:* 471–488 (1966).

Williams, M.A.; Smith, L.D.: Ultrastructure of the 'germinal plasm' during maturation and early cleavage in *Rana pipiens.* Devl Biol. *25:* 568–580 (1971).

Wilt, F.: The ontogeny of chick embryo hemoglobin. Proc. natn. Acad. Sci. USA *48:* 1582–1590 (1962).

Wittek, M.: La vitellogenèse chez les amphibiens. Archs Biol. *63:* 134–198 (1952).

Wolpert, L.: Positional information and the spatial pattern of cellular differentiation. J. theor. Biol. *25:* 1–47 (1969).

Wolpert, L.: Positional information and pattern formation. Curr. Top. Dev. Biol. *6:* 183–224 (1971).

Woodland, H.R.; Graham, C.F.: RNA synthesis during early development of the mouse. Nature, Lond. *221:* 327–332 (1969).

Wright, D.A.; Subtelny, S.: Nuclear and cytoplasmic contributions to dehydrogenase phenotypes in hybrid frog embryos. Devl Biol. *24:* 119–140 (1971).

Yamada, T.: Beeinflussung der Differenzierungsleistung des isolierten Mesoderms von Molchkeimen durch zugefügtes Chorda- und Neuralmaterial. Okajimas Folia Anat. Jap. *19:* 132–197 (1940).

Yamada, T.: Cellular synthetic activities in induction of tissue transformation; in De Reuck, Krugel, Cell differentiation. Ciba Fdn Symp., pp. 116–130 (Little, Brown, Boston 1967).

Zagris, N.: Erythroid cell differentiation in unincubated chick blastoderm in culture. J. Embryol. exp. Morph. *58:* 209–216 (1980).

Zagris, N.; Melton, C.G.: Hemoglobins in single chick erythrocytes as determined by a differential elution procedure. Z. Naturforsch. *33:* 330–336 (1978).

Zwilling, E.: Interaction between limb bud ectoderm and mesoderm in the chick embryo. II. Experimental limb duplication. J. exp. Zool. *132:* 173–187 (1956).

Zwilling, E.: Limb morphogenesis. Adv. Morphogen. *1:* 301–330 (1961).

Zwilling, E.; Hansborough, L.A.: Interaction between limb bud ectoderm and mesoderm in the chick embryo. III. Experiments with polydactylous limbs. J. exp. Zool. *132:* 219–239 (1956).

Index

This index is restricted to the most important terms. The numbers refer to the pages where the terms are defined or used for the first time. The asterisks indicate new terms or terms used in a sense different from the usual. Refer to figure 1 for terms of descriptive embryology.